Contributions to Management Science

The series *Contributions to Management Science* contains research publications in all fields of business and management science. These publications are primarily monographs and multiple author works containing new research results, and also feature selected conference-based publications are also considered. The focus of the series lies in presenting the development of latest theoretical and empirical research across different viewpoints.

This book series is indexed in Scopus.

Cecilia Grieco

Consumer Behavior in the Sharing Economy

Adoption, Engagement, and Post-Usage Dynamics

Springer

Cecilia Grieco
Sapienza University of Rome
Rome, Italy

ISSN 1431-1941 ISSN 2197-716X (electronic)
Contributions to Management Science
ISBN 978-3-031-76278-9 ISBN 978-3-031-76279-6 (eBook)
https://doi.org/10.1007/978-3-031-76279-6

This Springer imprint is published by the registered company Springer Nature Switzerland AG
The registered company address is: Gewerbestrasse 11, 6330 Cham, Switzerland

To my husband Gabriele, for constantly supporting my professional growth, and to our three beloved daughters, for doing their charming best to hinder it.

Acknowledgments

Just a few words, because at the end of this adventure I would like to express my sincere gratitude to all who have (willingly or unwillingly) shared it with me.

This book gathers the results of several years of research on the topic of the sharing economy yet has only come to fruition in the last few months. This would not have been possible without the support of valued professionals, my evergreen mentors, and the support and patience of my family.

Firstly, my Executive Editor, Dr. Prashanth Mahagaonkar, who has accompanied me with care and great professionalism. Once more he has seen the potential of my idea and helped me develop it at every stage to its final outcome. I hope there will be further opportunities for collaboration in the future.

Then, Dr. Mark Palmer, who worked with me on proofreading each chapter of this book. His competence and reliability have been pivotal for the completion of the project.

A special thank you is due to my mentor, Professor Gennaro Iasevoli, who encouraged me in this project and spent the whole summer reading the chapters and sending me feedback. His support has always been there throughout my career, and I hope he is aware of my deep gratitude to him.

Also, my unofficial mentors who are just as important, and not only from a professional point of view. Laura Michelini without whom I would not be here writing books during the summer. I owe her everything and I am always honoured to share my achievements with her. Francesca Ieracitano who, unfortunately for her, knows every line of my research and is always ready to listen to my complaints and encourage me to get back on board. I could not imagine doing this work without you.

Several people helped me indirectly, giving me the quiet and support I needed to focus on my project: my beloved family, my friends, my ballet mates. I am glad to have you in my life. Special thanks to Velia, who never fails to encourage me. I hope I will always be worthy of your pride.

Lastly, because the most important ones always come at the end, my husband Gabriele and my three little monkeys to whom this book is dedicated, even if I am pretty sure none of them will ever read it. Thank you for patiently sharing all the ups and downs of this project. I love you.

Contents

Chapter 1
Introduction

This could be as big as the Industrial Revolution in the way we think about ownership [Quoted in Sacks (2011)].

Rachel Botsman—Saïd Business School Oxford University

The importance of possession to contemporary society has been called into question in the last two decades. In Western countries in particular, ownership has been diminishing in significance, with the access and use of assets becoming more relevant than owning them (Rifkin, 2001). Recent years have also seen the growth in popularity of minimalist consumption, where experiences are valued more highly than material goods (Pangarkar et al., 2021; Morewedge et al., 2021). The sharing economy (hereafter SE) phenomenon intersects with both these societal and economic changes; for this reason, its development has led to substantial growth across industries and countries. At its core, SE indicates the possibility, provided by online platforms and digital technologies, for individuals to share products and services in return for various benefits. Hence it is configured as an alternative to ownership, allowing individuals to satisfy their needs while relieving them of the liability and burden of ownership. Furthermore, being largely based on the access-over-ownership principle, SE represents the transition from solidification (i.e. ownership-based consumption) to liquidity (Saravade et al., 2021). As the following chapters highlight, SE consists of a "varied range of old, revamped and new practices" (Schor & Fitzmaurice, 2015, p. 6). The practice of sharing is far from new, yet technological developments, increasing urbanization, and growing environmental awareness, as well as the financial constraints produced by the 2008 global economic crisis, have fostered the incorporation of this principle into a set of business models that are transforming the market for consumer goods and services, in addition to associated consumption practices (Schor & Fitzmaurice, 2015; Basselier et al., 2018).

The recession that followed the 2008 crisis led to a diffusion of initiatives aimed at offering the unused capacity of assets with a view to gaining, to a greater or lesser extent, additional revenue. Although this kind of initiative widely existed before this moment, increased financial demands coupled with the need to be frugal brought

C. Grieco, *Consumer Behavior in the Sharing Economy*, Contributions to Management Science, https://doi.org/10.1007/978-3-031-76279-6_1

about an impressive surge in sharing-based business models (Ranjbari et al., 2018). The goal was to find creative approaches capable of replacing traditional ownership-based consumption, motivated by the need to reduce customer costs. While initial attempts took place at a local level, mostly involving families and neighbours, the boost given by the advance of digitalization has spread these initiatives to a worldwide audience. This was the real innovation that has both determined the growing success of these platforms and seriously challenged incumbents while redefining industry boundaries in a range of businesses.

The SE phenomenon has been studied from different perspectives, including marketing (Lamberton & Rose, 2012; Kumar et al., 2018). Granting access to assets between providers and consumers has had disruptive implications for those traditional marketing frameworks dealing with market exchanges (Eckhardt et al., 2019). Within this field, studies on SE consumers have been plentiful (Merino-Saum et al., 2023). Since SE is largely based on peer-to-peer interactions, an in-depth understanding of consumer behaviour in sharing business models is particularly relevant. The underlying principles of SE business models profoundly reshape the overall journey of individuals within these platforms (Quinones & Augustine, 2015). For example, while traditional firms are based on the dyadic provider–consumer relationship, SE presents a triadic one built on the mediating role of the platform (Benoit et al., 2017). This complicates several aspects of the consumer journey, since, in all its stages, consumers are dealing with both platform and provider. Despite differences, these entities are closely related, with spillovers from one party to the other in terms of trust, disappointment, technical issues, and interactions.

Focusing on changes in consumer behaviour is strategic for advancing this field, both at academic and managerial levels. From the academic perspective, a recent work by the author (Grieco & Palagonia, 2024) illuminates the recent steady growth in contributions on SE consumers from 2017 onwards, consistent with studies on the SE phenomenon overall (De las Heras et al., 2021). This preliminary work resulted in the development of a model for consumer behaviour in SE, whose steps and dynamics require further investigation. The goal of this book is, within this context, to provide a systematization of a wide range of studies and to conceptualize the process of SE adoption, engagement, and post-usage.

From the managerial perspective, practitioners need to acquire a deep understanding of their consumers for two key reasons. The first is that, as traditional companies, SE platforms need to attract and retain consumers to survive. Although beyond the scope of this book, this need extends to both consumers and providers, for the success of SE platforms largely depends on their ability to exploit direct and indirect network effects that result from appropriate care being given to both sides. To have loyal consumers is the ultimate goal for SE platforms. Managers need to increase their knowledge of the motives driving consumers to SE platforms, the needs they expect to be satisfied, and the variables by which they evaluate both service quality and, consequently, their level of satisfaction. The second reason relates to the densely populated competitive scenario, where SE platforms need to compete with both traditional and other SE companies. From this point of view, SE practitioners need to undergo a two-step process to win consumers. The first challenge is to persuade individuals to move away from the traditional consumption pattern and to embrace the access philosophy; this implies a mindset shift which may be obstructed by several barriers. Consumers prone to value access over ownership are then able to choose from a wide range of platforms, given that the expansion of SE has resulted in intense competition (Zervas et al., 2017).

The book is structured into three chapters. Chapter 2—Framing the scope of the research. Definitions and boundaries of the sharing economy—presents the current state of the SE field of research, an essential step in defining the boundaries of the research and in identifying key definitions and perspectives within the topic. Despite being pivotal in introducing the topic of the book, defining the concept of SE is not an easy task. Also referred to as collaborative consumption, collaborative economy, or peer-to-peer economy, SE emerges as an umbrella concept, with multiple definitions that differ depending on their inclusiveness. The lack of a commonly agreed definition, probably due to the rapid development of the field, has led to the proliferation of terms used interchangeably by both practitioners and scholars. In confronting this confusion in the theoretical background, the aim of this chapter is to summarize and systematize the existing literature, focusing on the key elements of the SE concept and on the specificities differentiating this domain from any related concepts. Specific attention will be given to the impact SE has on several levels. This is paramount given how scholars line up either as enthusiastic supporters or as radical sceptics. The first group consists of those who stress the benefits SE can bring to society at different levels, starting from the core principles of optimizing the exploitation of underutilized resources, democratizing access to products and services, increasing social bonding, fostering micro-entrepreneurship, and encouraging measures to reduce ecological harm. In contrast, several authors point to the undesirable implications of SE activities, ranging from their actual green credentials to the exploitation of workers and the promotion of unsustainable consumption patterns. A final consideration in this chapter attends to the ecosystem of actors composing the business model of SE platforms. This derives from the nature of multi-sided markets which involve different kinds of actors in the value creation, delivery, and capture mechanisms.

The following two chapters focus on key stages in the SE consumer journey. Chapter 3—To share or not to share? Approaching the sharing economy—sets out the path leading consumers to adopt and use SE platforms, shedding light on the drivers, barriers, and key issues experienced by consumers. The branch of research into drivers is well established, with economic benefits often recognized as the key driver leading consumers to use SE platforms, regardless of sector. These platforms allow users to gain (providers) and save (consumers) money compared to traditional B2C transactions. A key role is also played by sustainability and the desire to adopt greener forms of consumption; sensitivity towards a more sustainable lifestyle is often a priority among sharing users. The search for the social experience of being part of a community is also a powerful driver, as well as social projection, enjoyment, and anti-consumption. A further element highlighted in this chapter is the trust issue. Peer-to-peer exchanges imply the creation of both economic and social bonding between users, and for this reason, it is essential that the platform adopts appropriate mechanisms to increase trust and, in turn, reduce potential hazards. An essential aspect also in traditional B2C transactions, what makes trust a delicate topic for SE is the fact that exchanges take place between users who do not know each other; for this reason, it can become the main barrier to user participation. Two further topics are discussed in this chapter: the segmentation schemes of SE consumers and the barriers that obstruct participation. Both aspects are pivotal in offering a complete understanding of SE consumers. This is particularly true for non-users, a topic that is poorly covered in the existing literature, a deeper understanding of which is very important for platform managers.

Finally, Chap. 4—"Happily ever after": The post-purchase behaviour of SE consumers—continues with the analysis of the SE consumer journey focusing on the post-purchase stage, when consumers form their perception of SE after having used it. From the platform perspective, this is a strategic phase, often facing several challenges to the effectiveness of retention strategies. SE has a transaction-centred nature that complicates the creation of switching costs and customer lock-in strategies. Moreover, the type of service itself often implies low transaction frequency, when business models address a market where product or service transactions occur infrequently. In this chapter, the discourse on post-purchase behaviour has two different levels of analysis. From a narrow perspective, it covers the considerations consumers have about their experience with the platform, such as the level of satisfaction, continuance intention, and loyalty. Despite enjoyment and economic benefits emerging as key elements, regardless of industry, studies into satisfaction in SE are often focused on specific countries or specific industries. As for continuance intention and loyalty, existing literature identifies the associated benefits for platforms, such as lower transaction costs, positive word of mouth, increased repeat purchases, and relational, social, and safety benefits for the users. Nevertheless, post-purchase behaviour will also be approached from a wider perspective, focusing on the extent to which satisfied consumers decide to adopt the access-over-ownership principle as a permanent consumption pattern, at least for specific categories of products and/or services. This phenomenon is called regime shift; it goes beyond the single transaction and implies a mindset shift. This aspect of consumption raises the issue of long-term consumer behaviour, drawing attention to the ownership concept and asking whether it is going to be superseded.

Each chapter is based on an extensive literature search and provides a detailed overview of the key contributions to the topics described above. The chapters are enhanced by thematic boxes which provide an analysis of specific SE platforms (Case study), data from empirical investigations (Insights), and snapshots of related key concepts (Focus).

References

Basselier, R., Langenus, G., & Walravens, L. (2018). The rise of the sharing economy. *Economic Review, 3*, 57–78.

Benoit, S., Baker, T. L., Bolton, R. N., Gruber, T., & Kandampully, J. (2017). A triadic framework for collaborative consumption (CC): Motives, activities and resources & capabilities of actors. *Journal of Business Research, 79*, 219–227.

De las Heras, A., Relinque-Medina, F., Zamora-Polo, F., & Luque-Sendra, A. (2021). Analysis of the evolution of the sharing economy towards sustainability. Trends and transformations of the concept. *Journal of Cleaner Production, 291*, 125227.

Eckhardt, G. M., Houston, M. B., Jiang, B., Lamberton, C., Rindfleisch, A., & Zervas, G. (2019). Marketing in the sharing economy. *Journal of Marketing, 83*(5), 5–27.

Grieco, C., & Palagonia, C. (2024). Delving into the behaviour of sharing economy consumers: A literature review. *Journal of Consumer Marketing, 41*(2), 162–179.

Kumar, V., Lahiri, A., & Dogan, O. B. (2018). A strategic framework for a profitable business model in the sharing economy. *Industrial Marketing Management, 69*, 147–160.

Lamberton, C. P., & Rose, R. L. (2012). When is ours better than mine? A framework for understanding and altering participation in commercial sharing systems. *Journal of Marketing, 76*(4), 109–125.

Merino-Saum, A., Jemio, P. R., Hansmann, R., & Binder, C. R. (2023). Drivers and barriers to participation in the sharing economy: Does the environment really matter? A systematic review of 175 scientific articles. *Resources, Conservation and Recycling, 198*, 107121.

Morewedge, C. K., Monga, A., Palmatier, R. W., Shu, S. B., & Small, D. A. (2021). Evolution of consumption: A psychological ownership framework. *Journal of Marketing, 85*(1), 196–218.

Pangarkar, A., Shukla, P., & Taylor, C. R. (2021). Minimalism in consumption: A typology and brand engagement strategies. *Journal of Business Research, 127*, 167–178.

Quinones, A., & Augustine, A. (2015). Technology and trust: How the sharing economy is changing consumer behavior. *US Banking Watch, 19*.

Ranjbari, M., Morales-Alonso, G., & Carrasco-Gallego, R. (2018). Conceptualizing the sharing economy through presenting a comprehensive framework. *Sustainability, 10*(7), 2336.

Rifkin, J. (2001). *The age of access: The new culture of hypercapitalism*. Penguin.

Sacks, D. (2011). The sharing economy, *Fast Company*. Retrieved from http://www.fastcompany.com/magazine/155/the-sharing-economy.html

Saravade, S., Felix, R., & Fırat, A. F. (2021). From solidity to liquidity: Macro-level consumption patterns in the sharing economy. *Journal of Macromarketing, 41*(2), 284–296.

Schor, J. B., & Fitzmaurice, C. J. (2015). Collaborating and connecting: The emergence of the sharing economy. In *Handbook of research on sustainable consumption* (pp. 410–425). Edward Elgar Publishing.

Zervas, G., Proserpio, D., & Byers, J. W. (2017). The rise of the sharing economy: Estimating the impact of Airbnb on the hotel industry. *Journal of Marketing Research, 54*(5), 687–705.

Chapter 2
Framing the Scope of the Research

Definitions and Boundaries of the Sharing Economy

2.1 The Defining Elements of the Sharing Economy

More than 10 years ago, Rachel Botsman, a pioneer of collaborative thinking, gave her article on the Fast Company the title "The sharing economy lacks a shared definition" (Botsman, 2013). One year later, Juliet Schor, a professor at Boston College, provided an unintended answer stating that "coming up with a solid definition of the sharing economy that reflects common usage is nearly impossible" (Schor, 2016, p. 2). Despite more than 10 years having since elapsed, these statements still sum up the output of academic and managerial efforts to suggest a consensual definition of the concept "sharing economy" (SE). To the general understanding, SE is an economic system based on people sharing possessions and services, either for free or for payment, usually organized through the Internet (Cambridge Dictionary[1]). Yet, a close look at academic research into the topic unveils a fragmented and nuanced interpretation of the concept, with each element of this definition widely conceptualized by scholars.

[1] Sharing economy: an economic system that is based on people sharing possessions and services, either for free or for payment, usually using the Internet to organize this. Source: https://dictionary.cambridge.org/dictionary/english/sharing-economy

C. Grieco, *Consumer Behavior in the Sharing Economy*, Contributions to Management Science, https://doi.org/10.1007/978-3-031-76279-6_2

2.1.1 The Role of the Digital

According to Rojanakit et al. (2022), there are two common elements across the definitions that capture the essentials of SE: connecting people through the Internet and the availability of resources and capabilities open to sharing. As for the former, the sharing phenomenon is largely rooted in digital transition, where the development of ICT is the main enabler of the rise and success of these platforms. This is an aspect on which consensus has been reached among scholars (Sutherland & Jarrahi, 2018), among whom "online platforms" is the most frequent term in SE definitions (Ranjbari et al., 2018). Digitalization has allowed the transition from the timeless practice of sharing among small groups of close individuals to the emergence of online platforms connecting a wider number of people sharing a larger pool of resources (Minami et al., 2021; Pouri & Hilty, 2021). Transitioning to digital technologies has freed sharing practices from their traditional limitations, ensuring an efficient and flexible exchange system within a virtually unlimited network of users.

In the sharing domain, the role of the platform is to ensure a match between suppliers and demanders while providing a convenient and secure environment for both parties. Several economic benefits accrue from the use of platforms: primarily, fixed costs are significantly lower than in conventional businesses, being mostly limited to establishing and maintaining the website or application (Zervas et al., 2017). Furthermore, the costs of human resources performing the core activities of the business are often close to zero, as the exchanges take place among consumers, and the community of users itself covers tasks that would otherwise incur expenses for the provider (Pouri & Hilty, 2021). Digital platforms also make it easier for users to join the community and participate in the sharing system both as consumers and as providers. This adds to the indirect network effects produced when the growth of one party increases value for the other, allowing for a solid scalability of the business (Wirtz et al., 2019).

Based on the results of an extensive literature analysis, Sutherland and Jarrahi (2018) clarify the role of technology in SE identifying six affordances for users. These affordances are defined as the productive relationships between people or groups that emerge from their interactions with the materiality of the SE platform. In this sense, the authors describe the role of technology in SE through its emergence from interactions (Sutherland & Jarrahi, 2018). The six affordances identified by the authors are (i) generating flexibility—accessing the platform is dynamic and rapid, and participants can take on different roles through online participation; (ii) match-making—the platform optimizes contact between suppliers and demanders; (iii) extending reach—given the wide scale, distance, and range of resources and individuals SE can bring together; (iv) managing transactions—the platform covers issues related to logistics, security, and payments; (v) trust building—by establishing a system to make exchanges reliable; and (vi) facilitating collectivity—the platform fosters and benefits from the collective action generated by SE (Sutherland & Jarrahi, 2018) (Table 2.1).

Table 2.1 The role of digital—Examples of definitions

Author/s (Year)	Who shares
Heinrichs (2013)	"These systems take a variety of forms ***but all leverage information technology***"
Barnes and Mattsson (2016)	"The use of onl***ine marketplaces and social networking technologies*** to facilitate peer-to-peer sharing of resources"
Filippas and Gramstad (2016)	"This marketplace structure, also referred to as online peer-to-peer marketplaces, ***has been made possible by a number of technological advances such as internet-enabled mobile devices***"
Pargman et al. (2016)	"The practice of making use of underutilized physical resources that had previously been neglected by ***using computer systems as a mediator***"

Source: Author's elaboration

2.1.2 Who Shares: The Parties Involved

A further agreed-upon aspect of SE is the opportunity provided to consume resources collaboratively (Ranjbari et al., 2018). This reflects what really makes SE innovative compared to historical practices of sharing, that is, sharing among strangers (Schor, 2016). While traditional sharing happened within families or close communities, the novelty of SE lies in the possibility of connecting with others beyond users' habitual social networks. The debate around this aspect concerns the nature of the asset owners and, consistently, the systems of exchange. Based on the traditional practice of peer-to-peer sharing, a subset of definitions sees SE as involving private individuals, both as providers as and consumers (Gerwe & Silva, 2020). This implies that those practices where businesses and professional operators act as providers fall outside the boundaries of SE. On the other hand, several authors open to the possibility of including B2C exchanges within SE also regard government and non-governmental organizations as potential users (Codagnone et al., 2016; Michelini et al., 2018). A third way is proposed by Ranjbari et al. (2018) and Markman et al. (2021). They get around the C2C versus B2C debate by proposing a definition where both forms are envisaged as long as both sides are independent from the platforms. In their view, businesses can participate in SE if they offer access to resources through a platform they do not own; otherwise we would be talking about a product-service system where the intermediary role of the platform is questioned (Ranjbari et al., 2018). Conversely, other scholars include equally within SE both companies that provide a platform and those that own the resources (Biswas et al., 2015; Roover, 2016) (Table 2.2).

2.1.3 What to Share: The Object of Exchange

Beyond the central role played by digital platforms and the involvement of individuals, questions remain about the other defining elements of the sharing economy. Focusing on the object of the exchange, scholars tend to refer to the unused capacity of tangible and/or intangible assets while presenting a broad range of

Table 2.2 Who shares—Examples of definitions

Author/s (Year)	Who shares	Perspective
Böckmann (2013)	"The term sharing economy refers to a business model where the participants share unused resources among them ***via a peer-to-peer services***"	Narrow
Carroll and Bellotti (2015)	"***Person-to-person exchanges*** of commodities and services (…) exchanges are arranged through social media interactions and are executed through face-to-face interactions"	
Barnes and Mattsson (2016)	"The use of online marketplaces and social networking technologies to facilitate peer-to-peer sharing of resources ***between individuals, who may be both suppliers and consumers***"	
Wang et al. (2019)	"Sharing economy as ***peer-to-peer interactions***, fee-based for customer with the aim of allowing to maximize under-utilized goods"	
Botsman (2013)	"It is currently largely talked about in relation to P2P marketplaces, but ***equal opportunity lies in the B2C models***"	Broad
Ranjbari et al. (2018)	"(…) in the peer-to-peer networks that it has created, because of the trust built among its members, ***who may be individuals or businesses***"	
Lim (2020)	"The sharing economy is a marketplace that consists of ***entities (e.g., consumers, organisations)*** that innovatively and sustainably shape how marketing exchanges of valuable products and resources are produced and consumed through sharing"	

Source: Author's elaboration

interpretations. Taking a narrower perspective, the object of the exchange consists in tangible goods that are shareable because by their nature they provide owners with excess capacity that can be lent out to others (Frenken & Schor, 2019). This happens with most consumer goods, such as houses, cars, clothes, and tools; this underscores the centrality of the unused capacity concept in defining SE. As reported by Belk (2014), some scholars suggest that shareable goods are "lumpy" (i.e. only available in discrete bundles regardless of users' needs) and "granular" (i.e. expensive, and thus only affordable through sharing).

Adopting a wider view, Pouri and Hilty (2021) refer to different degrees of shareability, depending on the materiality (versus immateriality) and on the durability (versus consumability) of the resource. With respect to durable material goods, the concept of unused capacity takes on a different meaning if increased use allows the object to fulfil more demand in parallel or in sequence. The first case refers to the concept of free capacity (e.g. occupying free seats in a car through car sharing) and the second to that of idle capacity (e.g. sharing a household tool among neighbours). Consumable material goods are, in contrast, exhausted when consumed and can be shared to avoid waste (e.g. food approaching its expiry date). With respect to immaterial goods, these are durable if they allow for repeated use without value degradation (e.g. software and skills) and consumable if volatile (e.g. information about traffic shared in real time). Although pivotal in many definitions, Markman et al. (2021) question the need to specify that the shareable capacity has to be unused. In

Table 2.3 What to share—Examples of definitions

Author/s (Year)	What to share	Perspective
Von Hoffen et al. (2015)	"(…) allow private individuals to ***access physical resources*** owned by other private peers for a limited period of time"	Narrow
Zervas et al. (2017)	"The emergence of peer-to-peer platforms, collectively known as the 'sharing economy,' has enabled individuals to collaboratively make use of ***underutilized inventory*** via fee-based sharing"	
Barnes and Mattsson (2016)	"The use of online marketplaces and social networking technologies to facilitate peer-to-peer sharing of resources (***such as space, money, goods, skills and services***)"	Broad
Hamari et al. (2016)	"A peer-to-peer-based activity of obtaining, giving, or sharing access to ***goods and services***, coordinated through community-based online services"	
Pouri and Hilty (2021)	"The digital sharing economy is a class of resource allocation systems based on sharing practices which are coordinated by digital online platforms and performed by individuals and possibly (non-) commercial organizations with the aim to provide access to ***material or immaterial resources***"	

Source: Author's elaboration

their view, this is a superfluous specification, as physical assets and human labour are almost always underutilized; thus, it is not something that is unlocked specifically by the sharing economy (Markman et al., 2021). What they stress in their definition is that the objects of the exchange have to be rivalrous in their use (i.e. that they cannot be accessed by more people at the same time) and not owned by the platform.

The key element that distinguishes narrow from wide perspectives on the objects of sharing is the extent to which the object of the exchange is a capacity that would otherwise be wasted versus a capacity that is generated for the purpose of being shared. Scholars who agree with the narrow perspective argue that creating new capacity is outside the scope of SE; it falls within the on-demand economy, based on peer-to-peer service delivery rather than on the peer-to-peer sharing of goods. On the other hand, scholars who adopt the wide perspective are open to the inclusion of the ad hoc provision of goods and services within the definition of SE (Frenken & Schor, 2019) (Table 2.3).

2.1.4 How to Share: The Ownership Issue

The discourse on the object of exchange runs parallel with the ownership issue and related considerations about the ways in which users share assets. Consistent with the interpretation of SE as an economic system that allows the shared use of resources, most definitions strictly exclude practices involving the transfer of ownership (Belk, 2007, 2014). The focus is thus on the temporary access granted to users (Frenken et al., 2015); no good is bought or sold but is collaboratively used

Table 2.4 How to share—Examples of definitions

Author/s (Year)	How to share	Perspective
Belk (2007)	"Sharing is an ***alternative to the private ownership*** that is emphasized in both marketplace exchange and gift giving. In sharing, two or more people may enjoy the benefits (or costs) that flow from possessing a thing"	Narrow
Benoit et al. (2017)	"No ownership transfer, often ***shared ownerships***, therefore simultaneous or sequential use"	
Ranjbari et al. (2018)	"Sharing and renting are also stated in many other definitions; however, selling, should not be considered in such systems, because ***the transfer of ownership does not happen in the SE, and a temporary access to the assets is provided*** for the consumers"	
Schor (2016)	"Sharing economy activities fall into four broad categories: ***recirculation of goods, increased utilization of durable assets, exchange of services, and sharing of productive assets***"	Broad
Mair and Reischauer (2017)	"A web of markets in which individuals use various forms of compensation ***to transact the redistribution of and access to resources***, mediated by a digital platform operated by an organization"	
Minami et al. (2021)	"Usually there is no transfer of ownership, ***unless it is in the form of giving***"	

Source: Author's elaboration

before being returned to the owner (Ranjbari et al., 2018). In fact, in its very first mention, the SE concept was denoted as the activity of sharing, exchanging, and renting of resources without owning the goods (Lessig, 2008). This strict view largely lies within the access-over-ownership principle in shaping how SE is conceived. Nevertheless, there is no unanimity among scholars, some of whom opt for a more inclusive interpretation that allows for the recirculation of goods (Mair & Reischauer, 2017; Rahim et al., 2017). An example comes from Schor (2016), for whom online marketplaces such as eBay and Craigslist are the first examples of the sharing economy and provide the origins of the phenomenon in the late 1990s. Some scholars also propose intermediate positions, such as Minami et al. (2021), who reiterate the absence of ownership transfer unless the exchange happens in the form of a donation (Table 2.4).

2.1.5 Why to Share: Compensation Mechanisms

A further aspect under debate is the form of compensation participants might expect to receive. Scholars' interpretations here are positioned on a continuum of possibilities reaching from the mere pleasure of community feeling to the explicit intention to earn money.

Pouri and Hilty (2021) address this issue when talking about the social aspect of sharing, stating how several possible levels of expected compensation for the access provided are envisaged in SE platforms. The authors recognize how legitimate expectations can be based on formal and/or informal reciprocity or compensation.

Authors such as John (2013) firmly believe that SE operates without money changing hands, as its goal is not enrichment. In the same vein, Belk (2014) critically labels as "pseudo-sharing" all those practices that masquerade as sharing in order to benefit from the positive associations coming from this definition while being determined by market mechanisms and involving revenue (e.g. long-term renting and leasing, short-term rental, online sites sharing data, online-facilitated barter economies). However, this view is contested by studies demonstrating that the possibility of economic gain is a powerful motivator for SE (see par. 2.1). In their literature review, Ranjbari et al. (2018) found that most definitions in their analysed sample insist on the profit motive of SE, or on some form of benefit in kind, even if not necessarily pecuniary. In this sense, practices such as gift-giving or charities are excluded from the SE domain (Table 2.5).

2.2 Two Perspectives on Sharing Economy

The overview here presented illuminates how scholars' perspectives on the definition of SE range from narrow to wide (Fig. 2.1). Narrow interpretations frame SE as having specific features in terms of ownership transfer, compensation, sharing

Table 2.5 Why to share—Examples of definitions

Author/s (Year)	Why to share	Perspective
Lessig (2008)	"***Non-price-based social relations*** act as a primary source of resource allocation in the sharing economy"	Narrow
Minami et al. (2021)	"Sharing economy consists of the practice of using and sharing products or services between two or more individuals with the support of the Web 2.0, and that ***does not involve any form of material compensation***"	
Botsman (2013)	"An economic model based on sharing underutilized assets from spaces to skills to stuff ***for monetary or non-monetary benefits***"	Broad
Frenken et al. (2015)	"Consumers granting each other temporary access to under-utilized physical assets ("idle capacity"), ***possibly for money***"	
Markman et al. (2021)	"Compensation to asset owners can ***take many forms, including financial payments, barter, and other forms of exchange***. Given that the overwhelming majority of platforms are digital, and payments often entail online transactions, qualifying the sharing economy based on whether transactions, exchanges, or access are conducted online or offline seems unhelpful"	

Source: Author's elaboration

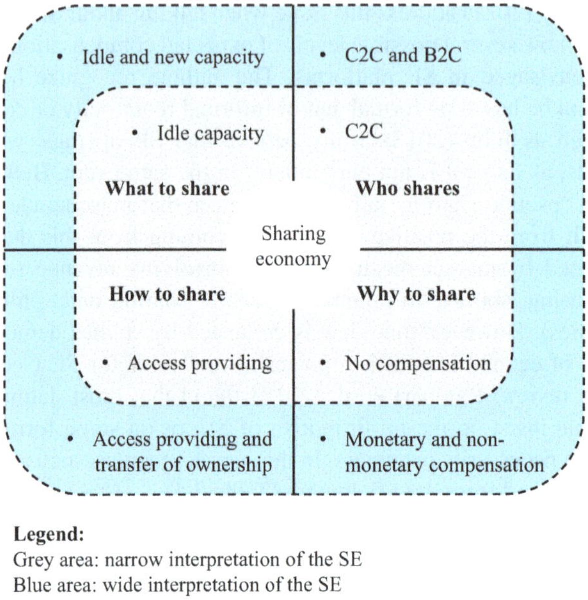

Fig. 2.1 Defining elements of sharing economy: two perspectives. Grey area: narrow interpretation of SE. Blue area: wide interpretation of SE. Source: Author's elaboration

modes, and the actors involved, proposing perspectives that might diverge. According to the wide perspective, the sharing economy can be referred to as an umbrella concept (Acquier et al., 2017; Habibi et al., 2017) that encompasses both C2C and B2C exchanges and resource access, but also service production, the reselling of goods, and different kinds of compensation mechanisms. This divergence both prevents us from identifying a consensual definition and further complicates our understanding of the phenomenon. As alluded to by Codagnone et al. (2016), a wide perspective tends to be over-inclusive to the extent that every digital platform could theoretically fall within the definition of SE, while narrow perspectives are under-inclusive and leave out platforms commonly considered to belong to SE.

It is worth noting that the two perspectives outlined are not mutually exclusive. In some cases, scholars provide definitions of SE which employ the narrow interpretation for some elements and the wide for others. Table 2.6 lists a set of the most cited definitions of SE (Schlagwein et al., 2020; Akbari et al., 2022; Khalek & Chakraborty, 2023) with radar graphs showing how, under the same definition, both interpretations can coexist.

Outlining the current state regarding the definition of SE has been essential in framing the scope of this investigation. In this book, a wide perspective is adopted to frame the SE phenomenon and to study the behaviour of its consumers. The reason behind this choice is twofold. First, definitions belonging to the narrow perspective often offer interpretations of SE that are highly divergent. This means that the

Table 2.6 Merging different interpretations within sharing economy definitions

Main definitions of SE	
"An economic model based on sharing underutilized assets from spaces to skills to stuff for monetary or non-monetary benefits. It is currently largely talked about in relation to P2P marketplaces, but equal opportunity lies in the B2C models"	
Botsman (2013), The Sharing Economy Lacks a Shared Definition, *Fast Company*	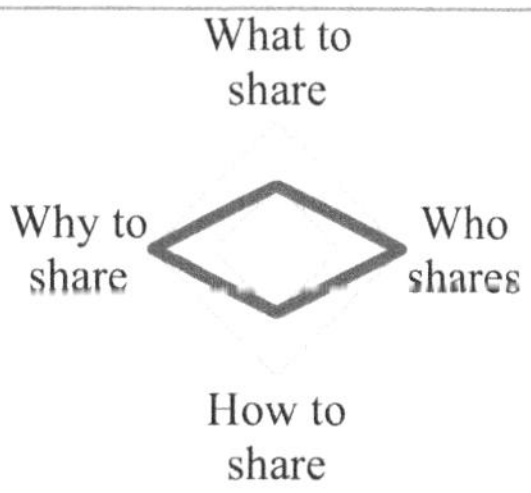
"Peer-to-peer-based activity of obtaining, giving, or sharing access to goods and services, coordinated through community-based online services"	
Hamari, J., Sjöklint, M., & Ukkonen, A. (2016), The sharing economy: Why people participate in collaborative consumption. *Journal of the association for information science and technology*	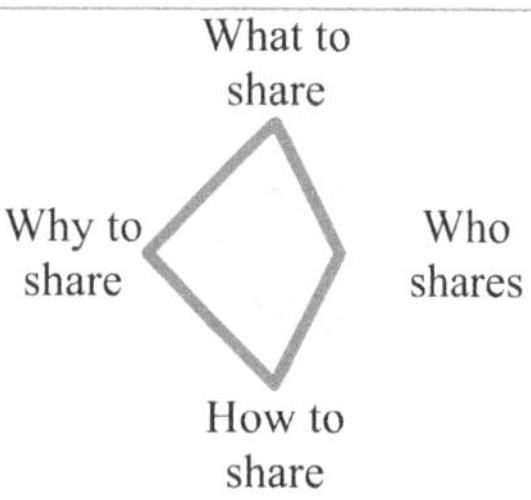
"Sharing economy describes the phenomenon as peer to peer sharing of access to under-utilised goods and services, which prioritizes utilization and accessibility over ownership, either for free or for a fee"	
Cheng (2016), Sharing economy: A review and agenda for future research. *International Journal of Hospitality Management*	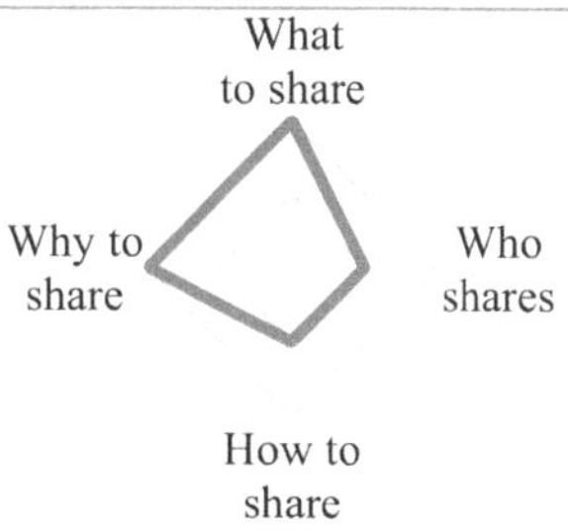
"Consumers granting each other temporary access to under-utilized physical assets ('idle capacity'), possibly for money"	

(continued)

Table 2.6 (continued)

Main definitions of SE	
Frenken and Schor (2019), Putting the sharing economy into perspective. *Environmental Innovation and Societal Transition*	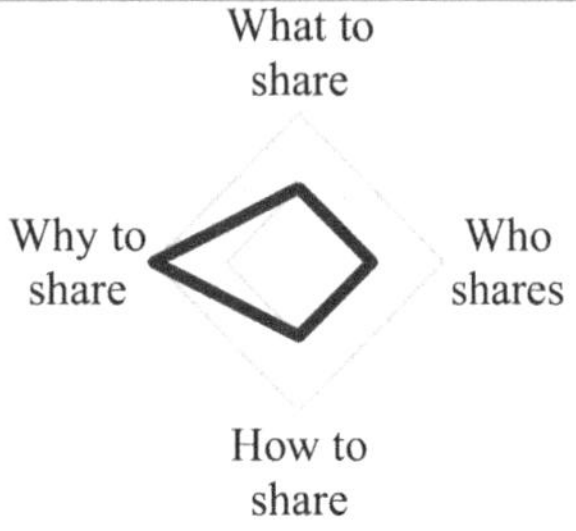
"A web of markets in which individuals use various forms of compensation to transact the redistribution of and access to resources, mediated by a digital platform operated by an organization"	
Mair and Reischauer (2017), Capturing the dynamics of the sharing economy: Institutional research on the plural forms and practices of sharing economy organizations, *Technological Forecasting and Social Change*	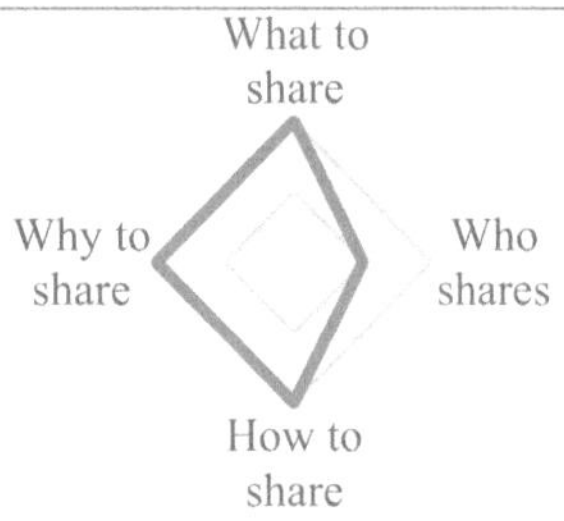
"An economic system, whose intermediary companies utilize online platforms to facilitate and lower the cost of the for-profit transactions of giving temporary access—without the transfer of ownership—to idle resources of consumers in the peer-to-peer networks that it has created, because of the trust built among its members, who may be individuals or businesses"	
Ranjbari, M., Morales-Alonso, G., & Carrasco-Gallego, R. (2018), Conceptualizing the sharing economy through presenting a comprehensive framework, *Sustainability*	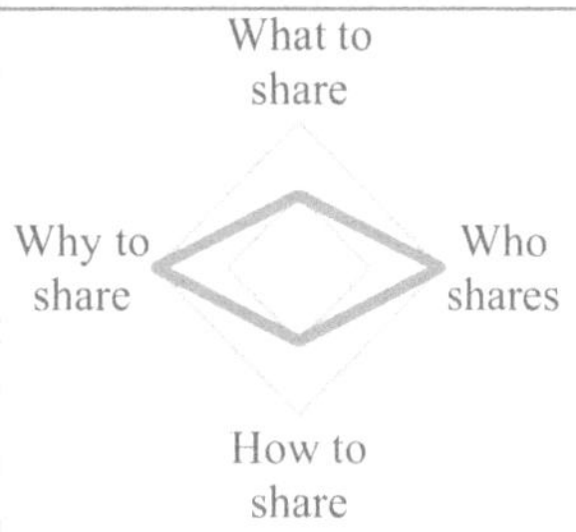
"Thus, we define the sharing economy as entailing users, asset providers, and multisided platforms that facilitate temporary access to –rather than ownership of –assets that are rivalrous in their use, and that are not owned by said platforms (…) Moreover, compensation to asset owners can take many forms, including financial payments, barter, and other forms of exchange"	

(continued)

Table 2.6 (continued)

Main definitions of SE	
Markman, G. D., Lieberman, M., Leiblein, M., Wei, L. Q., & Wang, Y. (2021), The distinctive domain of the sharing economy: definitions, value creation, and implications for research, *Journal of Management Studies*	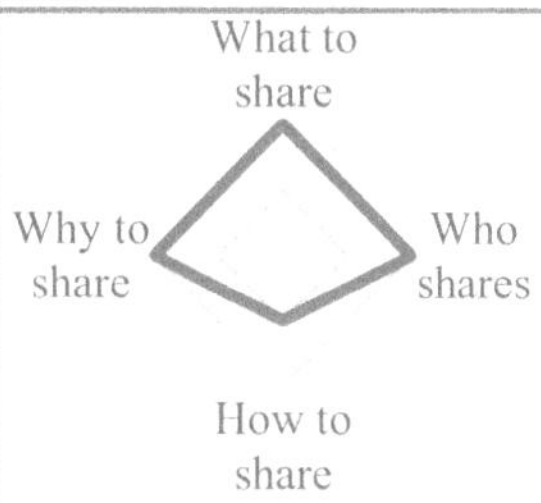
"The digital sharing economy is a class of resource allocation systems based on sharing practices which are coordinated by digital online platforms and performed by individuals and possibly (non-) commercial organizations with the aim to provide access to material or immaterial resources"	
Pouri and Hilty (2021), The digital sharing economy: A confluence of technical and social sharing, *Environmental Innovation and Societal Transitions*	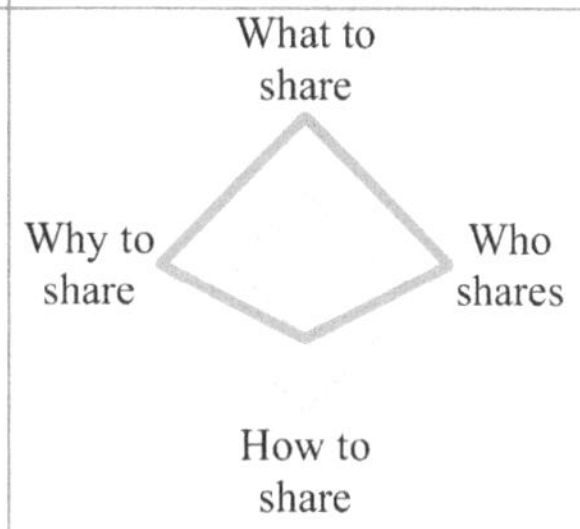

Source: Author's elaboration

same platform (e.g. Airbnb) is both included and excluded from SE based on different authors' accounts, further complicating any engagement with the topic. The second reason lies within the intention of this book to be of interest to both scholars and practitioners. Beyond the academic debate, managerial approaches to SE are influenced to a greater degree by the wide perspective, to the point that practices such as second-hand markets and co-working platforms are included as evolutionary forms of SE (Investopedia, 2024). For these reasons, and for the purpose of this book, SE is here understood as any exchanges taking place between individuals or businesses where a digital intermediary allows the renting, sharing, reselling, bartering, or gift-giving of assets for either monetary or non-monetary compensation.

2.3 Differences and Similarities Within the Related Phenomena

2.3.1 An Interpretative Framework

The lack of a commonly agreed definition of SE, probably due to the rapid development of the field (Hawlitschek et al., 2016), has led to the proliferation of terms that are used interchangeably in both the academic and professional domains, with scant

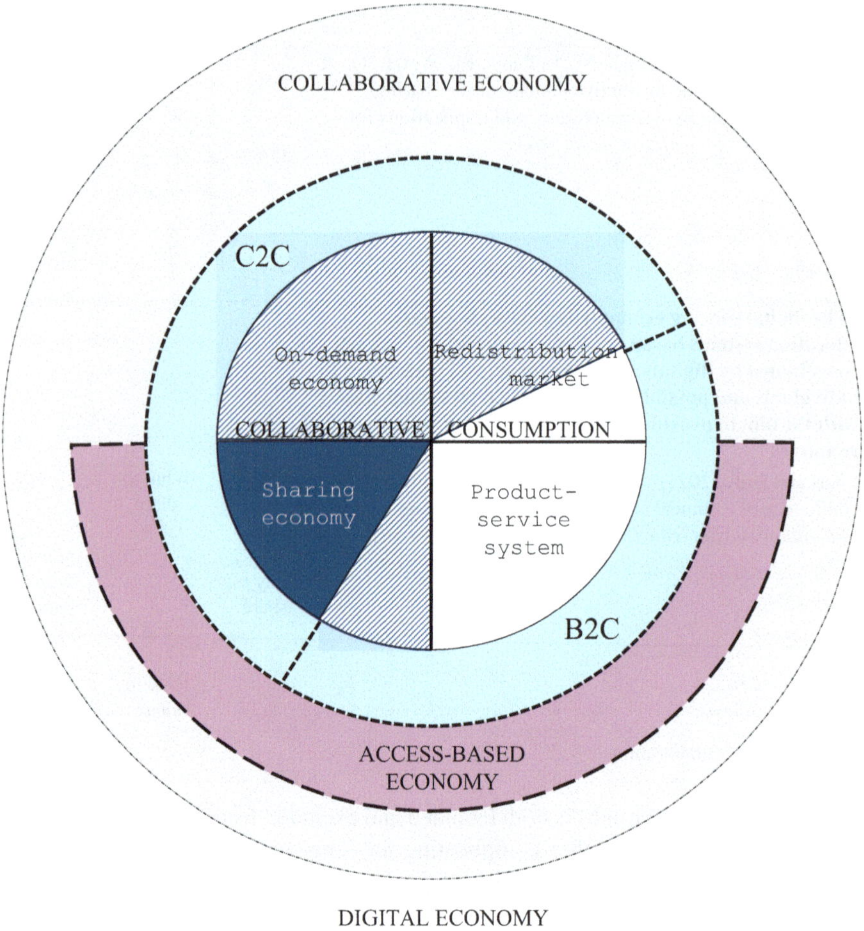

Fig. 2.2 Framing sharing economy among its related concepts. Blue area refers to the narrow interpretation of SE. Blue striped area refers to the wide interpretation of SE. Source: Author's elaboration

agreement over their boundaries. Scholars attempting to distinguish SE from related phenomena have discovered the extent to which the concept is deeply intertwined with others (Görög, 2018; Ranjbari et al., 2018; Sutherland & Jarrahi, 2018; Anwar, 2023), such as the peer-to-peer economy, collaborative economy, collaborative consumption, the product-service-system, the on-demand economy, gig economy, and platform economy. All of these emphasize a different aspect of an overall change to economic organization (UN, 2020). Despite some similarities, these practices differ greatly in other respects. Based on extant contributions, the purpose of this section is to clarify the main differences between the concepts while providing a framework for understanding the SE phenomenon which, given the confusion over its definition, appears controversial (Fig. 2.2).

As previously stated, being ICT-enabled, what makes the SE phenomenon innovative compared to the past is that it is deeply rooted in the digital economy. This term refers to a broad range of economic activities that (i) are produced through the collection and use of digitalized information and knowledge; (ii) happen within modern information networks; and (iii) are driven by the effective use of information and communication technology (Chen & Wang, 2019). According to the Bureau of Economic Analysis (BEA), the digital economy includes the digital-enabling infrastructure needed for a computer network to exist and operate, the digital transactions that take place using that system (e-commerce), and the content that digital economy users create and access (digital media), (Barefoot et al., 2018). All purchases and sales of goods and services that occur via computer networks are considered to belong to the e-commerce domain, as well as platform-enabled transactions, either between individuals or companies. This makes the so-called platform economy into a subset of the digital economy and indicates the tendency for commerce to develop and favour digital platform business models. This allows the creation of commercial ecosystems that are interactive and less hierarchically centralized than traditional firms (Bai & Velamuri, 2021). Scholars have identified different typologies of digital platforms based on their key elements (Markman et al., 2021), and SE may be associated with transaction platforms that allow people and organizations to share information or buy, sell, or access a variety of goods and services (Cusumano et al., 2019). Its distinctive features lie in the role of the platform as facilitating the process thanks to which asset providers and demanders are matched. In many instances, the emergence of the platform economy has replaced older business models, as is the case with video rentals, where video streaming services (e.g. Netflix) have replaced transactions hitherto largely carried out in physical stores. SE has also had disruptive effects in some industries, where rapid growth in its significance has made platforms such as Airbnb, Uber, and Lyft highly popular in a very short time span, thereby challenging traditional businesses that have taken much longer to become big players in their industries.

The digital economy and development of transaction platforms are at the core of the rise of the so-called access-based economy (or access economy), where the platforms are leveraged to develop business models that allow community members to rent assets temporarily in return for a fee, while the legal ownership remains with the service provider (Lawson et al., 2016). The access-based economy questions the actual need to own an asset, whose purchase can be cost-effectively replaced by providing easy to use and reliable access to it (Omazić et al., 2023). SE has much in common with the access-based economy, which, however, extends to the B2C (or even B2B) system, where the business itself provides both the platform and assets to exchange.

According to Botsman (2013), SE is part of collaborative consumption that is in turn a component of the wider collaborative economy, an economy built on networks of connected individuals and communities, including production, finance, and education. The economic model of the collaborative economy is built on the principle that ownership and access are shared between corporations, startups, and people, with a resulting market efficiency that brings new products, services, and business growth (Owyang et al., 2013). Within this paradigm, collaborative consumption allows the optimization of assets through efficient models of redistribution and shared access (Botsman, 2013). In its academic definitions, collaborative consumption appears very similar to the wide interpretation of SE. The first definition provided by Felson and Spaeth (1978) defined it as events in which individuals consume economic goods or services in the process of engaging in joint activities. The significance of the topic was raised by Botsman and Rogers (2010) who presented it as the practice of "sharing, bartering, lending, trading, renting, gifting, and swapping redefined through technology and peer communities" (Botsman & Rogers, 2010, p. xv). Given the wide number of practices included, it is possible to identify categories among the collaborative consumption phenomena. Botsman (2013) and Frenken and Schor (2019) point to three different systems: (i) redistribution markets (or second-hand economy) where consumers sell goods to each other allowing for ownership transfer instead of temporary access; (ii) on-demand economy (or gig economy) where consumers exchange skills and services instead of goods; (iii) product-service system where the exchanged asset is owned by a business provider which grants consumers long- or short-term access. In its narrow interpretation, SE is configured as a fourth system of collaborative consumption, characterized by consumers sharing temporary access to the idle capacity of assets without compensation. In contrast, taking the wide perspective, SE almost completely overlaps with collaborative consumption, except for the product-service system, where the asset provider is also the owner of the platform (Fig. 2.3).

2.3.2 *The Sharing Economy or Collaborative Consumption?*

Among the aforementioned related concepts, SE and collaborative consumption are the most widely quoted terms in the literature (Nguyen & Llosa, 2018). Yet, while there is general agreement over the meaning of related terms, such as access-based, peer-to-peer, or platform economy, the distinction between SE and collaborative consumption sees divergences among authors. This is because, as stated in the previous section, SE can be interpreted in different ways, leading to greater or lesser overlap with the concept of collaborative consumption.

Scholars adopting a wide perspective on SE tend to think it includes collaborative consumption. For example, Hamari et al. (2016) describe SE as a contemporary technology-driven phenomenon encompassing sub-concepts such as collaborative consumption. Huber (2017) states that collaborative consumption, together with access-based consumption practices, collective knowledge creation, and collaborative findings, is the main constituent of SE. Rinne (2017) considers that the two terms synonymously indicate the collaborative forms of consumption, production, finance, and learning. Similarly, Anwar (2023) uses the two terms referring to the same phenomenon based on their entanglement.

In contrast, a narrow interpretation leads authors to include SE in the wider phenomenon of collaborative consumption. Based on the analysis of 26 different related terms, Görög (2018) concludes that SE is narrower than collaborative consumption, which includes collaborative forms of consumption, production, finance, and

Sharing economy ≠

Digital Economy: encompasses those economic activities happening because of ICT, allowing for the development of SE.

Platform Economy: a subset of the digital economy based on digital transactions; a precondition of SE where the platform plays the central role of enabling the supplier-demander matching process.

Access-based Economy: the economic paradigm based on accessing assets instead of owning them. It partially overlaps with SE but goes further, involving B2C and B2B exchange systems.

Collaborative Economy: the economic paradigm based on the joint ownership and access of assets between market actors. SE is part of the collaborative economy a having the same underpinning principle.

Collaborative consumption: a component of the collaborative economy referring to sharing, bartering, lending, trading, renting, gifting and swapping, redefined through technology and peer communities. It coincides with the wide interpretation of SE.

Redistribution markets/second-hand economy: consumers sell goods to each other allowing for ownership transfer instead of temporary access. It differs from the narrow interpretation of SE that does not allow for ownership transfer, while being included in the wide perspective.

On demand economy/gig economy: consumers exchange skills and services instead of goods. It differs from the narrow interpretation of SE where the object of the exchange is restricted to goods, while being included in the wide perspective.

Product-service system: the exchanged asset is owned by a business provider who grants consumers long or short-term access.

Source: Author's elaboration

Fig. 2.3 How sharing economy differs from its related concepts. Source: Author's elaboration

learning, while the former only envisages the reuse of underutilized assets. Khalek and Chakraborty (2023) use nine defining variables of exchange to propose a taxonomy aimed at disentangling the different types of SE. What they come up with is the identification of collaborative consumption as a macro-category encompassing altruistic sharing, true sharing, and complementary sharing. At the same time, they refer to SE as the overall system of exchange encompassing both collaborative consumption and access-based consumption.

Given this semantic confusion, some contributions focus on the identification of the key differences between these concepts. Benoit et al. (2017) use three key features to differentiate SE from collaborative consumption and from more traditional forms of exchange. Firstly, collaborative consumption happens within a triadic relationship with the platform providers allowing the exchange; a customer in need of an asset; and a provider delivering the core service. In contrast, SE might include a platform but more usually occurs between individuals. Secondly, both SE and collaborative consumption do not envisage the transfer of ownership; rather, the provider grants access to the shared asset. The difference lies in the type of use granted that in SE can be simultaneous or sequential, whereas it is only sequential in collaborative consumption. Thirdly, the exchange is mediated by market mechanisms in collaborative consumption and by social mechanisms in SE. Focusing on the underlying predominant logic, Minami et al. (2021) shift attention to the consumer's perspective and highlight how these forms of consumption are associated with different core motivations. SE is underpinned by concern for the well-being of others; consistently it emerged that it is mostly driven by intrinsic reasons such as enjoyment and social and environmental orientation. Conversely, collaborative consumption is a self-centred form of consumption, with the key motivation driving consumers being the search for compensation, in the form of economic benefits, cost, or financial savings, and for their own satisfaction and positive customer experience.

The result of this confusion is that the same platform could be included in or excluded from SE basing on the perspectives of the different authors. For example, Airbnb is one of the most well-known platforms for short-term house rental. Searching for SE examples on the web would certainly lead to Airbnb, as it is generally considered among the main cases of the phenomenon. Yet, focusing on scholars' divergent definitions its belonging to the category cannot be taken for granted. According to Böcker and Meelen (2017) and Hamari et al. (2016), Airbnb is an example of SE as it implies consumers granting each other temporary access to their underutilized physical assets possibly for money. Schor (2016) considers Airbnb as practically synonymous of SE. Basing on Belk's (2014) view, Airbnb falls within the collaborative consumption concept as it includes individuals coordinating acquisition and distribution of resources for a fee. For the same reason, the author excludes Couchsurfing that is similar but does not foresee compensation in return for hospitality and thus it has to be included in the SE domain (Belk, 2014). The same is stated by Benoit et al. (2017) and Minami et al. (2021) who place Couchsurfing in the SE realm as there is no market mediation, while Airbnb falls under the collaborative consumption label for the opposite reason. Authors such as Gerwe and Silva (2020) exclude Airbnb from SE as in their view it requires that both the providers and the consumers are private individuals. In other cases, the platform is included only in the case of spare asset. This means that if the house owners rent their space while away for holidays it is renting its temporary idle capacity and thus it could be included in SE. Otherwise, those who buy a second house appositively to rent it out are actually running a commercial lodging site, even if it happens through the Airbnb platform and thus it cannot be considered as SE (Ranjbari et al., 2018; Frenken & Schor, 2019). The same confusion emerges for other platforms such as Uber, Lyft, or Blablacar. As proposed by Schor (2016) the best way to come out from this impasse is to follow the ways platforms define themselves and are defined by the press.

Insights: Sharing Economy and Pro-environmental Behaviour

The relationship between SE and sustainability is much debated. On the one hand, SE is praised as a sustainable solution for its potential to reduce consumption and recirculate and redistribute resources which would otherwise be discarded or left inactive. Yet, the extent to which SE consumers turn to SE platforms specifically because of their economic concerns has been questioned and, in some cases, disconfirmed.

The following figures present the results of an empirical investigation into the relationship between SE and pro-environmental behaviour, that is, the adoption of behaviours that harm the natural environment as little as possible or contribute to its protection (Hartig et al., 2001). Using the scale developed by Markle (2013), a sample of 128 active SE users were investigated with a view to understanding the extent to which they are also likely to adopt a pro-environmental lifestyle.

Conservation (1/2)

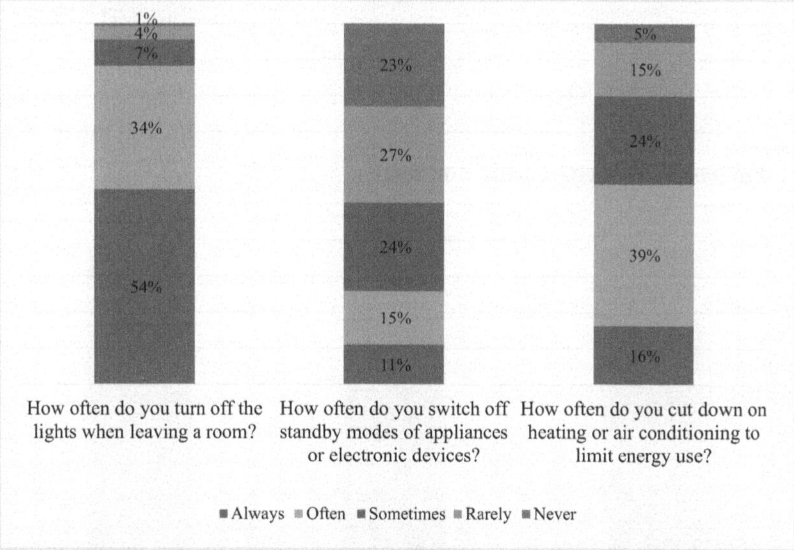

The scale investigates pro-environmental behaviour focusing on four areas: conservation, environmental citizenship, food, and transportation. As for conservation, some behaviours are more diffused than others. For example, nearly all respondents say they turn off the lights when leaving a room (54% always; 34% often) or wait for a full load before using the washing machine or dishwasher (40% always; 33% often), yet only a small proportion switch off appliances or electronic devices (11% always; 15% often) or limit the time in the shower to save water (13% always; 22% often).

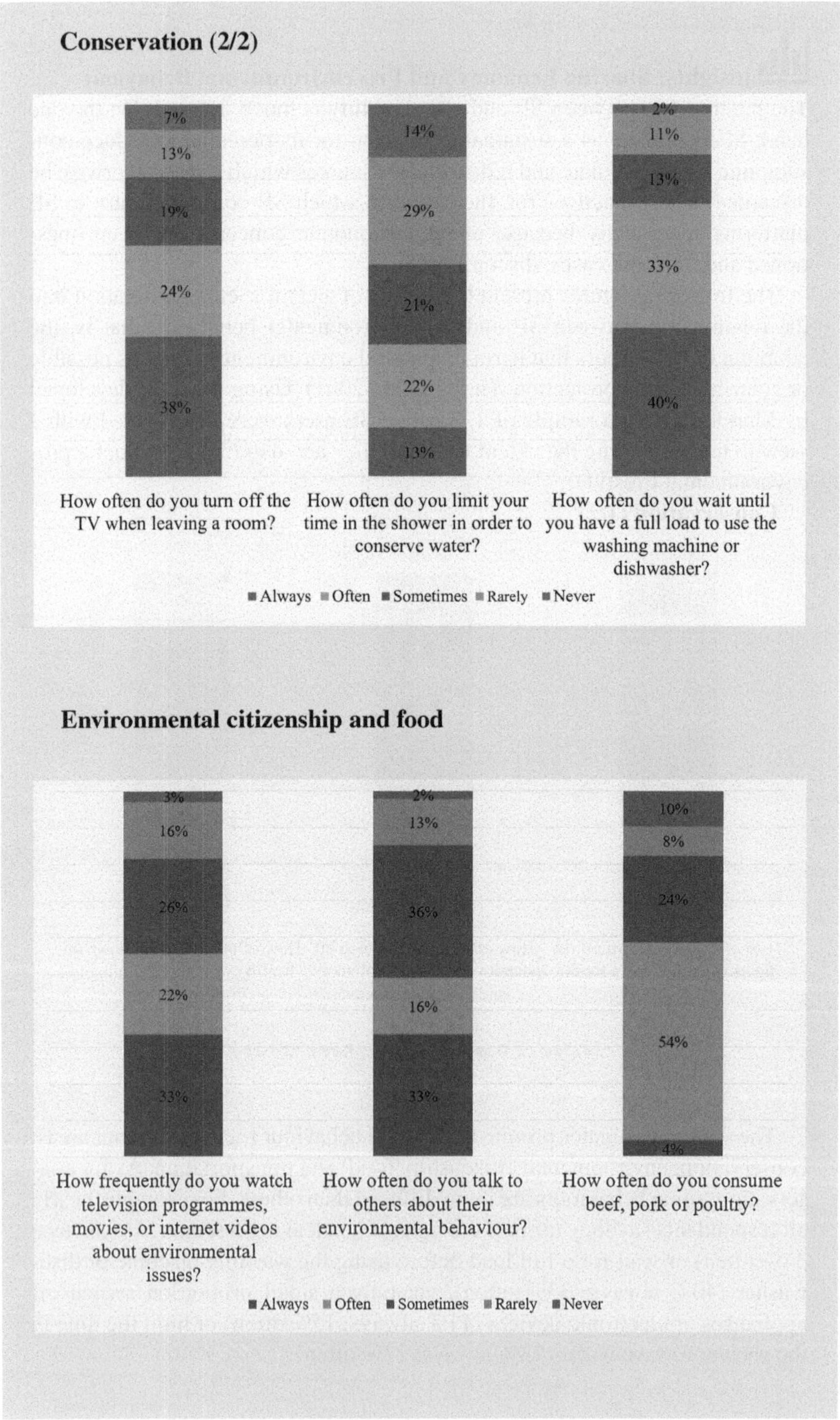
Conservation (2/2)
7%
13%
19%
24%
38%
How often do you turn off the TV when leaving a room?
14%
29%
21%
22%
13%
How often do you limit your time in the shower in order to conserve water?
2%
11%
13%
33%
40%
How often do you wait until you have a full load to use the washing machine or dishwasher?
Always Often Sometimes Rarely Never
Environmental citizenship and food
3%
16%
26%
22%
33%
How frequently do you watch television programmes, movies, or internet videos about environmental issues?
2%
13%
36%
16%
33%
How often do you talk to others about their environmental behaviour?
10%
8%
24%
54%
4%
How often do you consume beef, pork or poultry?
Always Often Sometimes Rarely Never

As for environmental citizenship, 33% of respondents say they always adopt an active behaviour, while for meat consumption, more than half the sample say they eat it 3–5 days a week (54%), with 4% having it every day

Transportation

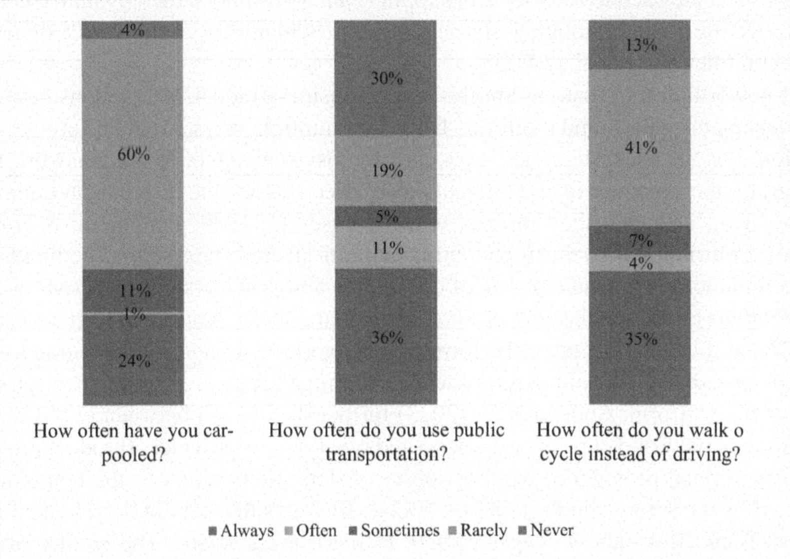

Questions about transportation are those with the highest margin of intervention. Car-pooling is a rarely used option, as is walking or cycling instead of driving. Public transportation is never used by 30% of respondents, against 36% who use it daily.

2.4 The Sharing Economy Ecosystem

Common across the different perspectives outlined above is the fact that SE refers to an exchange model that connects suppliers and demanders through the facilitation of an intermediary. SE is an example of a multi-actor business model that, in contrast to traditional businesses, operates through a triadic system of relations between customers, service providers, and platforms (Benoit et al., 2017).

The role of a certain type of technology intermediary is present in most studies on SE (Schlagwein et al., 2020). In some of these definitions, the role of the intermediary actor is not explicit yet implied by referring to the "marketplace" or "match-making", i.e. mediation facilitating the connection between those offering products and services and those needing to access them. As previously mentioned, the type of these two categories of actors varies depending on how narrowly or widely SE is interpreted. In the former (narrow) only peers are involved; the latter (wide) also envisages businesses (or non-profit and government organizations) on the supply side. In the narrow definition, the terms "people", "individuals", "peers", and "users" are used to refer to both suppliers and consumers, thereby underscoring how any user can simultaneously act as both (Schlagwein et al., 2020) or move between roles (Ertz et al., 2022).

It is worth noting that, within this triadic business model, interactions between customers, providers, and platforms happen in multiple ways. We therefore need to explore the SE ecosystem. The concept of ecosystem refers to a framework that integrates the existence of and interactions between actors, the external environment they act in, and who they interact with (Rothschild, 1990; Peltoniemi & Vuori, 2004). Contributions to this topic outline a multi-layered ecosystem. The first layer is on the micro level and consists of consumers and providers whose activities and interactions constitute the core of SE (Leung et al., 2019). According to Barari et al. (2022), at this level SE takes the form of a network of strangers exchanging assets to gain an expected benefit in return while avoiding the perceived risks related to the use of the platform. Abdalla et al. (2023) further distinguish between two types of consumers (new and current customers) and two types of providers (product providers and service providers), while a subset of definitions refers to the community formed by the actors, highlighting the social structures that extend beyond the mere transactions (Botsman & Rogers, 2010; Hamari et al., 2016). The quality of the consumer–provider relationship at the micro level predicts that of the relationship each party will develop with the platform, which, at the meso level, acts as a mediator, thanks to the assistance of modern technology. Costello and Reczek (2020) provide a comparative investigation of the relationship consumers form with providers and with the platform. In contrast to traditional businesses, consumers perceive providers as entities that are independent from the platform, opening the latter to different marketing and communication strategies. The authors conclude that provider-focused marketing communications are more effective in driving consumers to purchase, compared to platform-focused communications. This is because in the first case, what the authors call an "empathy lens" leads consumers to see the purchase as an act to help an individual rather than just a pecuniary transaction in return for a good or service (the "exchange lens") (Costello & Reczek, 2020).

Leung et al. (2019) identify a third layer, the exosystem; this includes those actors with the potential to influence and be influenced by the inner part of the system: partners that collaborate with the platform, competitors with traditional business models, and government as well as industry organizations and politicians. At the macro level, the last layer includes contextual variables (e.g. gross domestic product—GDP), cultural values, belief systems, and lifestyles (Leung et al., 2019; Barari et al., 2022).

Focus: The Ecosystem of Food Sharing Platforms

Food sharing platforms are a particular subset of SE and are characterized by the hybrid mission of generating a positive effect on the environment and society while ensuring they are economically and financially sustainable (Acquier et al., 2019). These platforms are able to partially mitigate the food waste issue by connecting waste generators and potential receivers (Ciulli et al., 2020). Examples of these platforms include TooGoodToGo, Olio, Regusto, and Bring the Food. Despite being similar in their mission, these platforms adopt different business models and thus involve different actors in their business models, leveraging their assets to increase their impact. Research by Michelini et al. (2023) divides the actors in the food sharing platform ecosystem into two main categories: (i) platform players are those directly involved in the value chain platform process and (ii) business players are those influencing the development and growth of the platforms by acting at social, political, economic, and cultural levels. The first category includes suppliers and beneficiaries, whether businesses, consumers, or NGOs, municipalities, and complementary assets providers (e.g. software developers, commercial partners). The second category includes volunteers, universities, local authorities, media, government institutions, and incubators (Michelini et al., 2023).

2.5 All the Shades of Sharing Economy: An Analysis of Negative Impacts

The rise of SE has been welcomed enthusiastically by different quarters (individuals, policymakers, and scholars) praising the potential benefits of these platforms for society. Arguments in favour of this economic paradigm include the promotion of communal forms of consumption and the development of trust, solidarity, and social bonding among participants (Benkler, 2017; Palgan et al., 2017). From the consumer perspective, SE platforms allow for low-cost transactions, flexible information exchange, and increased access to unused resources (Behl et al., 2022; Bucher et al., 2018; Pignot, 2023). From the provider perspective, the assistance of online platforms provides instant access to market information, ensures the best match with demanders, and facilitates interaction with consumers (Pan & Qiu, 2022). SE has also been seen as a stimulus for entrepreneurship (Botsman & Rogers, 2010) and as a way of overcoming the economic recession and growing environmental problems caused by consumption (Botsman & Rogers, 2010).

Yet, along with the positive assessments of SE, its development over the years has also brought a trail of scepticism in its wake. Following on from Belk's (2014) definition of pseudo-sharing, the main concern is the inclusion of the financial aspect and market mediation, whose inner logics nullify the positive nature of the sharing principle (Rana et al., 2023). According to some scholars, SE hangs in a precarious balance between the pursuit of financial objectives and the declared intent to safeguard the welfare of communities, where the mutually exclusive nature of these goals inevitably leads platforms to lean more towards the former than the latter.

In recent years, interest in its negative impact has seen a branch of studies emerge focused on the so-called dark side of SE, defined as the socially, environmentally, and economically undesirable effects produced by the introduction of the sharing practice (Buhalis et al., 2020). The impact of SE can be assessed on different levels. According to the triple bottom line perspective, SE has the potential to impact on the economy with cost reductions and a better use of resources; on society with interactions between members of communities; and on the environment by promoting shared consumption using fewer resources. Nevertheless, it has been seen as a source of value destruction in all three areas, as well as there being scant evidence to support its sustainability claims (Schor, 2016), with the risk of promoting increased consumption (Parguel et al., 2017), and the threat to privacy and labour rights (Vith et al., 2019). Within the SE ecosystem, different types of interactions between actors can potentially generate both positive and negative effects at micro, meso, and macro levels (Fig. 2.4).

Scholars agree on the role of the platform as crucial in determining (and mitigating) the negative effects of SE (Wu & Benson-Rea, 2024). A first reason relates to the inappropriate internal management processes that lead to the bankruptcy of many platforms after their fast entry and rapid expansion in the market (Si et al., 2021). Despite being attracted by the opportunity to surf the wave of

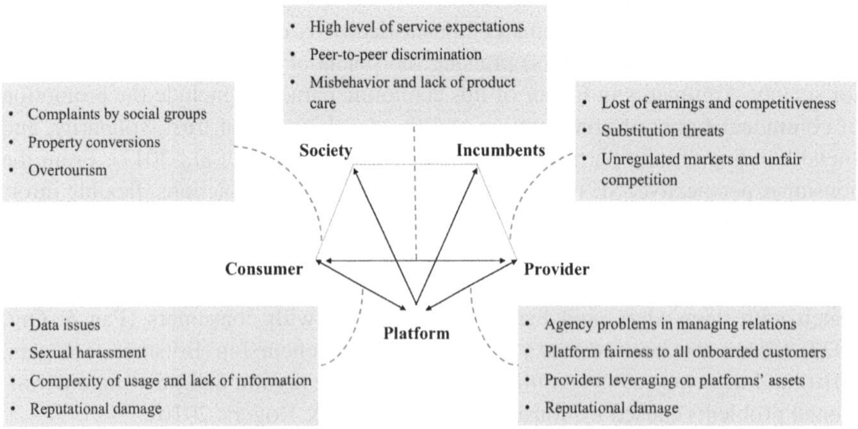

Fig. 2.4 The negative impacts of sharing economy. Source: Author's elaboration

the sharing phenomenon, not all the players can implement a successful management system capable of facing changes in market requirements as well as operational risks (Wu & Benson-Rea, 2024). As multisided markets, SE platforms can benefit from indirect network effects coming from nurturing equally both suppliers and demanders. However, this implies the ability of the platform to keep both sides satisfied and to take responsibility for damages caused by one side to the other. The misbehaviour of providers may result in reputational loss for the platforms, as was the case in 2018 with Uber, which received over 3000 claims of sexual assault, raising serious concerns about safety and accountability (Bellon, 2019). This is a unique feature of the SE business model, where customers interact with different service providers, and their (dis-)satisfaction determines the quality of their relationship with the platform. Similarly, providers' experiences with customers will determine whether they stay with the platform (Barari et al., 2022).

Impact on providers comes from both their internal management and business model decisions and the external relations with the platform and the end users (Wu & Benson-Rea, 2024). As for the former, SE offers the opportunity to gain income and to interact with customers in an immediate and personal way (Osman, D'Acunto). On the other hand, it often brings to the marketplace players who are not always well-equipped to carry out an entrepreneurial activity and struggle to find successful business strategies. Providers are often micro-entrepreneurs without the requisite capabilities to compete in the market; rather, they leverage on the platform's strategic assets, leading to operational uncertainty (Wilhelms et al., 2017). The relationship with the platform can be problematic, most of all because of the power disparity that puts them in the position of pressing their advantage over the providers and maximizing extraction of value created. In this context, the main concern is the unclear employment status of workers in SE, their unfair treatment, scarce or even non-existent earnings, and lack of social protection (UN, 2020). This stems from the large bargaining power of platforms which can impose critical barriers to access the service through non-inclusive algorithmic rules set by the platform, or to exit when providers want to terminate their contract (Rana et al., 2023). Provider discontent has often resulted in heated protests, such as the strikes by Lyft and Uber drivers demanding better pay and working conditions that are worse than traditional employees (O'Brien, 2019). Despite the platforms responding with evidence that their average pay levels are above the minimum wage for drivers, data show how, from 2019 to 2022, passengers' fares increased at a faster rate than drivers' pay (Zhou, 2024). Protests also aim to make platform managers aware of the lack of protection against workplace discrimination and unsafe working conditions providers experience in countries where they are not legally considered as employees. A study by Newlands et al. (2018) clustered a cross-national sample of 6111 SE providers based on their attitude towards collective action and providers' self-classification, showing how a substantial proportion of participants welcome unionization, consider it feasible, and take part in online communities. Yet, the majority would prefer to be classified as independent contractors and not as employees, thereby limiting organized forms of collective action. The emergence of these complexities within SE echoes a more general trend with platform workers, where the International Labour Organization (ILO) reported 1271 instances of protest from 2017 to 2020, mostly due to pay concerns, employment status, and health and safety. These instances cover strikes, logoffs, and demonstrations and are often coordinated by workers themselves and trade unions, with ride-hailing and food delivery being the principal platform sectors in which these events tend to happen (Bessa et al., 2022).

Focusing on the relationship with the end users, negative impacts come from their misbehaviour and a lack of care they might have for the shared assets, resulting in maintenance and cleaning costs. Product care allows for the extension of a product's lifetime and includes careful handling and protective measures to maintain its highest value (Ackermann & Tunn, 2024). As SE involves the shared use of the same products, product care becomes a very important issue. Product lifetime extension also contributes to sustainability. Yet what emerges is that consumers are more likely to take care of products they own and feel low psychological ownership in access-based consumption (Bardhi & Eckhardt, 2012). As reported by Ackermann and Tunn (2024), several studies raised the issue of deficient product care with access across different product categories such as cars (Bardhi & Eckhardt, 2012), strollers (Sumter et al., 2018), and washing machines (Ackermann & Tunn, 2024). This tendency to misbehave may create a snowball effect where the previous misbehaviour of consumers influences other consumers' behaviour (Schaefers et al., 2016). Providers are additionally subject to increased pressure to keep up with a high level of service expectations, further fuelled by rating systems and the need to achieve and maintain high scores (Buhalis et al., 2020).

From the consumer perspective, the key issue relates to personal data and uncertainty about their use. SE platforms are often accused of unclear data collection and treatment, with no explicit request for consent or transparent notification of the risks relating to privacy and security (Lu & Yi, 2023). A further area of potential risk for consumers is the complexity of usage coupled with the lack of information that impacts negatively on consumer satisfaction and discourages them from repurchase (Wu & Benson-Rea, 2024). In their analysis of grey literature, Rana et al. (2023) collected instances of consumers left in unsafe conditions while using car sharing services, with no support provided by the platforms. This was even more critical in emerging economies, where the cars had been described as dirty, unhygienic, and unsuitable for use (Rana et al., 2023).

SE has social implications and impacts on the well-being of both individuals and communities (Anderson et al., 2013; Breidbach & Brodie, 2017). Yet this impact can be highly negative when the transactions between users and providers cause problems and damage that redounds to third parties. The accommodation industry offers examples of neighbours' complaints due to nuisance and feeling endangered by home-sharing services, leading municipalities to regulate these platforms more strictly (Frenken & Schor, 2019). In many large cities such as San Francisco, Tokyo, and Vienna, the period when a property can be listed for short-term rental has been limited to 90 or 180 days (Financial Times, 2024). A very recent episode involved the citizens of Barcelona, Spain, who protested Airbnb and the short-term rentals it enables with the consequent impact on housing costs. The response of the city mayor was to commit to reducing short-term tourist letting licences in the next 4 years and thereby to return 10,000 apartments to the residential market (Financial Times, 2024).

Finally, SE has a major impact on incumbents (i.e. established firms with a traditional business model), to the point that some industries have been severely disrupted by its rapid development. This is the case in the hospitality and mobility industries where SE platforms (e.g. Airbnb and Uber) have attracted unconventional participants to the market, increasing competition while causing a significant decline in the earnings of traditional competitors (Zervas et al., 2017). The consequences of this have been tough for the employees of traditional firms, such as the taxi drivers who have lost ground in almost every city since the launch of car-sharing services (Zeng & He, 2023). In their analysis of online news articles, Leung et al. (2019) found that incumbents are the actors in the ecosystem most concerned by the rise and growth of SE, given the devastating effects it has had in cutting their competitiveness. Traditional competitors have reacted to these externalities through lawsuits and campaigns to protect their rights against newcomers; at the same time, they have found appropriate solutions to innovate their business models and remain competitive (Ciulli & Kolk, 2019).

A further impact on incumbents results from many unregistered businesses participating in SE, exploiting governance gaps to avoid paying taxes on their extra income as many existing regulations do not apply to the new industry. This contributes to the creation of an unregulated marketplace and unfair competition with traditional businesses (Wu & Benson-Rea, 2024).

Case Study: How Airbnb Impacted on the Accommodation Industry

Airbnb is undoubtedly the most referenced platform when speaking about SE in both academic and, even more so, managerial domains. The platform was founded in 2007, when Brian Chesky and Joe Gebbia decided to rent out air mattresses in their apartment in San Francisco to attendees of a conference, as all the hotels were booked. For this reason, they called the service "Air Bed and Breakfast", starting an experiment that, despite initial scepticism, grew impressively in the following years. Currently, Airbnb listings span over more than 200 countries and regions, with over five million hosts welcoming over 1.5 billion guests around the globe. As a multisided platform, Airbnb's business model connects hosts offering unused or underutilized spaces and guests looking for a place to stay for a short period of time. Its success is due to the compelling value proposition that offers economic benefits for both the parties while leveraging travellers' search for experience by offering them authentic local experiences.

However, the growth of this platform has been severely criticized for its negative externalities, leading to the definition of Airbnb as a disruptor of the accommodation industry (Hati et al., 2021). Following the model in Paragraph 1.4, Airbnb's both positive and negative impacts can be described according to the relationships with the other actors of its relevant ecosystem (Table 2.7).

Table 2.7 Positive and negative impact of Airbnb on its relevant ecosystem

	Positive impact	Negative impact
Providers	-Gaining additional revenues from rentals -Social interactions with the guests	- Customer-oriented approach reducing hosts' autonomy - Growing professionalization of the service expected
Consumers	-Authentic local experiences -Cost-effective solutions	- Actual service does not live up to the promised value - Safety and security concerns
Society	-City beautification processes -Interaction with tourist -Business opportunities in residential areas	- Gentrification - Reduced availability for long-term rentals - House price increases - Noise and overcrowding
Incumbents	-Inspiring business model innovation for traditional hotels	- Substitution effect for budget hotels/motels - Price increase in high-end hotels

Source: Author's elaboration

Impact on providers: Airbnb's hosts are the most powerful and influential stakeholders on the platform together with its guests. This is evidently because the overall business model relies on hosts' assets without which Airbnb could not provide its services. The positive effects of their involvement mainly consist in the possibility of having additional income by making use of otherwise unused spaces. Moreover, they can benefit from the social interaction with their guests (Buhalis et al., 2020). However, negative effects also arise, mostly linked to the power imbalance between the parties. In fact, Airbnb adopts a customer-oriented approach that in turn impacts on hosts' autonomy. Farmaki and Kaniadakis (2020) identify two related mechanisms by which Airbnb limits hosts' decision-making power. The first is the inclusion of emerging professionals who are more economically oriented and tend to offer services more in line with those of traditional hotels. This implies a gradual professionalization of the platform that causes an increase in guests' expectations and demand, even towards co-habiting hosts who have a more informal style of hosting that might include the guests contributing to cleaning the property (Farmaki & Kaniadakis, 2020). The same happens through Airbnb's changing policies in accordance with guest' preferences, for example, those limiting the host's power to select guests. This results in a growing pressure on hosts to stay competitive.

Impact on consumers: Airbnb guests are at the core of the platform's business model. The value creation process depends on both guests and hosts yet, as mentioned above, consumers' expectations guide the overall development of the platform. The benefits guests receive from the platform mostly refer to the opportunity to enjoying living like locals cost-effectively (Guttentag et al., 2019). Moreover, guest reviews are given greater weight by the platform compared to those of hosts, whose status is determined by the consumers' ratings (Gunter, 2018). Negative effects are generated when the platform does not act as a guarantor of consumers' expectations, for example, by allowing listings that do not accurately represent the accommodation (Birinci et al., 2018). Consumers might experience disappointment and dissatisfaction when the promised value does not correspond to the one provided, and the lack of professionalization of many hosts means that false descriptions, fake photos, overbooking, and cancellations happen more often than in traditional hotels (Buhalis et al., 2020). This could also end up in fraud and, in the worst cases, sexual harassment by hosts on guests (Buhalis et al., 2020).

Impact on society: Airbnb listings are usually concentrated close to the city centre and in proximity to the cities' points of interest, such as restaurants and pubs, attractions, retail, sports and entertainment, town halls, post-offices, and beaches, whose closeness has a positive impact on increasing the listing price (Quattrone et al., 2018). The presence of Airbnb is appreciated by residents who have the opportunity to interact with tourists, reduce loneliness, and develop business and job opportunities (Hati et al., 2021). The diffusion of the platform has brought interesting improvements to local areas, mostly improving the quality of life and promoting the physical rehabilitation of private buildings (Buhalis et al., 2020). Conversely, there are also negative effects on the quietness of neighbourhoods and the inflation the platform has triggered. As it relies on private properties, Airbnb has led to a mixing of residential and second home areas with visitor areas, merging the interests and needs of residents and tourists which are not always aligned (Hoffman & Heisler, 2020). The presence of listings means noise and pollution during the day and at night, as well as traffic and parking problems. Moreover, crime and antisocial behaviour have been reported and in the most serious cases including prostitution and drug issues (Manning et al., 2018). One of the most important aspects is the link between Airbnb and gentrification, i.e. the process by which poor neighbourhoods are gradually improved by the advent of middle-class residents replacing the working-class (Hammel, 2009). Gentrification is a controversial phenomenon. Although it can bring positive outcomes (e.g. an increase in economic activity and access to new urban facilities and services), its negative consequences include a reduction in housing options for vulnerable people, a worsening of social inequalities, displacement of local businesses, and increasing conflict between new and old residents. Improvements to an area may lead to an increase in property prices and, consequently, residential rental prices while reducing the availability of properties for long-term rent (Rabiei-Dastjerdi et al., 2022). Airbnb is certainly not the only element in this process; in some areas gentrification has been registered before the diffusion of the Airbnb platform. Yet, its contribution to neighbourhood change is significant (Rabiei-Dastjerdi et al., 2022).

Impact on incumbents: The main reason why Airbnb has been defined as a disruptive innovation is for the impact it had on both the tourism industry and, in particular, traditional hotels. Consistent with the developmental path of disruptive innovations (Christensen et al., 2013), Airbnb started by serving a niche market neglected by traditional accommodation services (Dolnicar, 2019). Then its popularity increased due to several factors, including the possibility for travellers to live like locals; the flexibility offered; the user-friendly website; the few assets it owns as properties belonging to individual providers; and the possibility of operating outside the government regulations that traditional hotels have to observe (Hati et al., 2021). Nevertheless, the impact of Airbnb in this context was uneven across the industry. As reported by Zervas et al. (2017), lower-end hotels and hotels not focused on business travellers have been the most affected, while upscale hotels are less so. Airbnb thus competes with traditional hotels on cost-benefit effectiveness, offering local experiences at an affordable price. This makes the competition quite a bit fiercer for incumbents who need to significantly reduce their prices to offer comparable overall value (Yang et al., 2019). Therefore, hotels serving high-end consumers have responded to the competition by going in an opposite direction and setting higher prices in those areas where the penetration of Airbnb is greater. This strategy is a result of the impossibility of competing with Airbnb on pricing while maintaining the same high service. These incumbents therefore prefer to move away from price-conscious consumers and concentrate more on their core segments; they signal their high-quality service by limiting discounts and deals (Roma et al., 2019).

References

Abdalla, S., Amankwah-Amoah, J., & Badewi, A. (2023). Sharing-economy ecosystem: A comprehensive review and future research directions. *Sustainability, 15*(3), 2145.

Ackermann, L., & Tunn, V. S. (2024). Careless product use in access-based services: A rebound effect and how to address it. *Journal of Business Research, 177*, 114643.

Acquier, A., Daudigeos, T., & Pinkse, J. (2017). Promises and paradoxes of the sharing economy: An organizing framework. *Technological Forecasting and Social Change, 125*, 1–10.

Acquier, A., Carbone, V., & Massé, D. (2019). How to create value (s) in the sharing economy: Business models, scalability, and sustainability. *Technology Innovation Management Review, 9*(2).

Akbari, M., Foroudi, P., Khodayari, M., Fashami, R. Z., & Shahriari, E. (2022). Sharing your assets: A holistic review of sharing economy. *Journal of Business Research, 140*, 604–625.

Anderson, L., Ostrom, A. L., Corus, C., Fisk, R. P., Gallan, A. S., Giraldo, M., et al. (2013). Transformative service research: An agenda for the future. *Journal of Business Research, 66*(8), 1203–1210.

Anwar, S. T. (2023). The sharing economy and collaborative consumption: Strategic issues and global entrepreneurial opportunities. *Journal of International Entrepreneurship, 21*(1), 60–88.

Bai, G., & Velamuri, S. R. (2021). Contextualizing the sharing economy. *Journal of Management Studies, 58*(4), 977–1001.

Barari, M., Paul, J., Ross, M., Thaichon, S., & Surachartkumtonkun, J. (2022). Relationships among actors within the sharing economy: Meta-analytics review. *International Journal of Hospitality Management, 103*, 103215.

Bardhi, F., & Eckhardt, G. M. (2012). Access-based consumption: The case of car sharing. *Journal of Consumer Research, 39*(4), 881–898.

Barefoot, K., Curtis, D., Jolliff, W., Nicholson, J. R., & Omohundro, R. (2018). *Defining and measuring the digital economy* (Vol. 15, p. 210). US Department of Commerce Bureau of Economic Analysis.

Barnes, S. J., & Mattsson, J. (2016). Understanding current and future issues in collaborative consumption: A four-stage Delphi study. *Technological Forecasting and Social Change, 104*, 200–211.

Behl, A., Jayawardena, N., Pereira, V., Tarba, S., & Bamel, U. (2022). The role played by relational turbulence in managing agency problems among value chain partners in the sharing economy: A review of the antecedents, benefits, risks, and boundary conditions. *Industrial Marketing Management, 107*, 39–51.

Belk, R. (2007). Why not share rather than own? *The Annals of the American Academy of Political and Social Science, 611*(1), 126–140.

Belk, R. (2014). Sharing versus pseudo-sharing in Web 2.0. *The Anthropologist, 18*(1), 7–23.

Bellon, T. (2019). Uber says it received over 3,000 reports of sexual assault in U.S. in 2018, *Reuters*. December 6, 2019, Retrieved from: www.reuters.com/article/us-uber-safety-idUSKBN1YA009

Benkler, Y. (2017). Peer production, the commons, and the future of the firm. *Strategic Organization, 15*(2), 264–274.

Benoit, S., Baker, T. L., Bolton, R. N., Gruber, T., & Kandampully, J. (2017). A triadic framework for collaborative consumption (CC): Motives, activities and resources & capabilities of actors. *Journal of Business Research, 79*, 219–227.

Berg, L., Slettemeås, D., Kjørstad, I., & Rosenberg, T. G. (2020). Trust and the don't-want-to-complain bias in peer-to-peer platform markets. *International Journal of Consumer Studies, 44*(3), 220–231.

Bessa, I., Joyce, S., Neumann, D., Stuart, M., Trappmann, V., & Umney, C. (2022). *A global analysis of worker protest in digital labour platforms*, ILO Working Paper 70 (Geneva, ILO).

Birinci, H., Berezina, K., & Cobanoglu, C. (2018). Comparing customer perceptions of hotel and peer-to-peer accommodation advantages and disadvantages. *International Journal of Contemporary Hospitality Management, 30*(2), 1190–1210.

Biswas, R., Pahwa, A., & Sheth, M. (2015). *The rise of the sharing economy: The Indian landscape*. Delhi.

Böcker, L., & Meelen, T. (2017). Sharing for people, planet or profit? Analysing motivations for intended sharing economy participation. *Environmental Innovation and Societal Transitions, 23*, 28–39.

Böckmann, M. (2013). The sharing economy: It is time to start caring about sharing; value creating factors in the sharing economy. Paper presented at *the 1st IBA bachelor thesis conference*, Enschede.

Botsman R. (2013). The sharing economy lacks a shared definition. *Fast Company*. Retrieved from https://www.fastcompany.com/3022028/the-sharing-economy-lacks-a-shared-definition#4

Botsman, R., & Rogers, R. (2010). *What's mine is yours: The rise of collaborative consumption*. Harper Business.

Breidbach, C. F., & Brodie, R. J. (2017). Engagement platforms in the sharing economy: Conceptual foundations and research directions. *Journal of Service Theory and Practice, 27*(4), 761–777.

Bucher, E., Fieseler, C., Fleck, M., & Lutz, C. (2018). Authenticity and the sharing economy. *Academy of Management Discoveries, 4*(3), 294–313.

Buhalis, D., Andreu, L., & Gnoth, J. (2020). The dark side of the sharing economy: Balancing value co-creation and value co-destruction. *Psychology and Marketing, 37*(5), 689–704.

Carroll, J. M., & Bellotti, V. (2015). Creating value together: The emerging design space of peer-to-peer currency and exchange. In *Proceedings of the 18th ACM conference on computer supported cooperative work & social computing* (pp. 1500–1510).

Chen, Y., & Wang, L. (2019). Commentary: Marketing and the sharing economy: Digital economy and emerging market challenges. *Journal of Marketing, 83*(5), 28–31.

Cheng, M. (2016). Sharing economy: A review and agenda for future research. *International Journal of Hospitality Management, 57*, 60–70.

Christensen, C., Raynor, M. E., & McDonald, R. (2013). *Disruptive innovation*. Harvard Business Review.

Ciulli, F., & Kolk, A. (2019). Incumbents and business model innovation for the sharing economy: Implications for sustainability. *Journal of Cleaner Production, 214*, 995–1010.

Ciulli, F., Kolk, A., & Boe-Lillegraven, S. (2020). Circularity brokers: Digital platform organizations and waste recovery in food supply chains. *Journal of Business Ethics, 167*, 299–331.

Codagnone, C., Biagi, F., & Abadie, F. (2016). *The passions and the interests: Unpacking the 'sharing economy'*. Institute for Prospective Technological Studies, JRC Science for Policy Report.

Costello, J. P., & Reczek, R. W. (2020). Providers versus platforms: Marketing communications in the sharing economy. *Journal of Marketing, 84*(6), 22–38.

Cusumano, M. A., Gawer, A., & Yoffie, D. B. (2019). *The business of platforms: Strategy in the age of digital competition, innovation, and power* (Vol. 320). Harper Business.

Dolnicar, S. (2019). A review of research into paid online peer-to-peer accommodation: Launching the annals of tourism research curated collection on peer-to-peer accommodation. *Annals of Tourism Research, 75*, 248–264.

Ertz, M., Boily, É., Sun, S., & Sarigöllü, E. (2022). Role transitions at the prosumer level: Spillover effects in the collaborative economy from an interactive marketing perspective. *European Journal of Marketing, 56*(10), 2721–2748.

Farmaki, A., & Kaniadakis, A. (2020). Power dynamics in peer-to-peer accommodation: Insights from Airbnb hosts. *International Journal of Hospitality Management, 89*, 102571.

Felson, M., & Spaeth, J. L. (1978). Community structure and collaborative consumption: A routine activity approach. *American Behavioral Scientist, 21*(4), 614–624.

Filippas, A., & Gramstad, A. R. (2016). A model of pricing in the sharing economy: Pricing dynamics with awareness-generating adoptions. In *Icis*.

Financial Times. (2024). *Barcelona and the Airbnb blacklash*. Retrieved from https://www.ft.com/content/287c1d53-7dd0-410c-88bb-f43277c851b6.

Frenken, K., & Schor, J. (2019). Putting the sharing economy into perspective. In *A research agenda for sustainable consumption governance* (pp. 121–135). Edward Elgar Publishing.

Frenken, K., Meelen, T., Arets, M., & Van de Glind, P. (2015). Smarter regulation for the sharing economy. *The Guardian*. Retrieved from https://www.theguardian.com/science/political-science/2015/may/20/smarter-regulation-for-the-sharing-economy

Gerwe, O., & Silva, R. (2020). Clarifying the sharing economy: Conceptualization, typology, antecedents, and effects. *Academy of Management Perspectives, 34*(1), 65–96.

Görög, G. (2018). The definitions of sharing economy: A systematic literature review. *Management* (18544223), 13(2).

Gunter, U. (2018). What makes an Airbnb host a superhost? Empirical evidence from San Francisco and the Bay Area. *Tourism Management, 66*, 26–37.

Guttentag, D. (2019). Progress on Airbnb: a literature review. Journal of Hospitality and Tourism Technology, 10(4), 814–844.

Habibi, M. R., Davidson, A., & Laroche, M. (2017). What managers should know about the sharing economy. *Business Horizons, 60*(1), 113–121.

Hamari, J., Sjöklint, M., & Ukkonen, A. (2016). The sharing economy: Why people participate in collaborative consumption. *Journal of the Association for Information Science and Technology, 67*(9), 2047–2059.

Hammel, D. J. (2009). Gentrification. *International encyclopedia of human geography* (pp. 360–367).

Hartig, T., Kaiser, F. G., & Bowler, P. A. (2001). Psychological restoration in nature as a positive motivation for ecological behavior. *Environment and Behavior, 33*(4), 590–607.

Hati, S. R. H., Balqiah, T. E., Hananto, A., & Yuliati, E. (2021). A decade of systematic literature review on Airbnb: The sharing economy from a multiple stakeholder perspective. *Heliyon, 7*(10).

Hawlitschek, F., Teubner, T., & Gimpel, H. (2016, January). Understanding the sharing economy--Drivers and impediments for participation in peer-to-peer rental. In *2016 49th Hawaii international conference on system sciences (HICSS)* (pp. 4782–4791). IEEE.

Heinrichs, H. (2013). Sharing economy: A potential new pathway to sustainability. *Gaia, 22*(4), 228–231.

Hoffman, L. M., & Heisler, B. S. (2020). *Airbnb, short-term rentals and the future of housing*. Routledge.

Huber, A. (2017). Theorising the dynamics of collaborative consumption practices: A comparison of peer-to-peer accommodation and cohousing. *Environmental Innovation and Societal Transitions, 23*, 53–69.

Investopedia. (2024). https://www.investopedia.com/terms/s/sharing-economy.asp

John, N. A. (2013). Sharing and Web 2.0: The emergence of a keyword. *New Media & Society, 15*(2), 167–182.

Khalek, S. A., & Chakraborty, A. (2023). Access or collaboration? A typology of sharing economy. *Technological Forecasting and Social Change, 186*, 122121.

Lawson, S. J., Gleim, M. R., Perren, R., & Hwang, J. (2016). Freedom from ownership: An exploration of access-based consumption. *Journal of Business Research, 69*(8), 2615–2623.

Lessig, L. (2008). Remix: Making art and commerce thrive in the hybrid economy. *Penguin*. https://doi.org/10.5040/9781849662505

Leung, X. Y., Xue, L., & Wen, H. (2019). Framing the sharing economy: Toward a sustainable ecosystem. *Tourism Management, 71*, 44–53.

Li, Z., Liang, C., Hong, Y., & Zhang, Z. (2022). How do on-demand ridesharing services affect traffic congestion? The moderating role of urban compactness. *Production and Operations Management, 31*(1), 239–258.

Lim, W. M. (2020). The sharing economy: A marketing perspective. *Australasian Marketing Journal, 28*(3), 4–13.

Lu, B., & Yi, X. (2023). Institutional trust and repurchase intention in the sharing economy: The moderating roles of information privacy concerns and security concerns. *Journal of Retailing and Consumer Services, 73*, 103327.

Mair, J., & Reischauer, G. (2017). Capturing the dynamics of the sharing economy: Institutional research on the plural forms and practices of sharing economy organizations. *Technological Forecasting and Social Change, 125*, 11–20.

Manning, C., deRoos, J., O'Neill, J. W., Bloom, B. A., Agarwal, A., & Roulac, S. (2018). Hotel/lodging real estate industry trends and innovations. *Journal of Real Estate Literature, 26*(1), 13–41.

Markle, G. L. (2013). Pro-environmental behavior: Does it matter how it's measured? Development and validation of the pro-environmental behavior scale (PEBS). *Human Ecology, 41*, 905–914.

Markman, G. D., Lieberman, M., Leiblein, M., Wei, L. Q., & Wang, Y. (2021). The distinctive domain of the sharing economy: Definitions, value creation, and implications for research. *Journal of Management Studies, 58*(4), 927–948.

Michelini, L., Principato, L., & Iasevoli, G. (2018). Understanding food sharing models to tackle sustainability challenges. *Ecological Economics, 145*, 205–217.

Michelini, L., Grieco, C., & Dentchev, N. (2023). Leveraging collaborations to increase the impact of food sharing platforms. *British Food Journal, 125*(11), 3953–3978.

Minami, A. L., Ramos, C., & Bortoluzzo, A. B. (2021). Sharing economy versus collaborative consumption: What drives consumers in the new forms of exchange? *Journal of Business Research, 128*, 124–137.

Muñoz, P., & Cohen, B. (2017). Mapping out the sharing economy: A configurational approach to sharing business modeling. *Technological Forecasting and Social Change, 125*, 21–37.

Newlands, G., Lutz, C., & Fieseler, C. (2018). Collective action and provider classification in the sharing economy. *New Technology, Work and Employment, 33*(3), 250–267.

Nguyen, S., & Llosa, S. (2018). On the difficulty to define the Sharing Economy and Collaborative Consumption-Literature review and proposing a different approach with the introduction of 'Collaborative Services'. *Journée de la Relation à la Marque dans un Monde Connecté*, 19–25.

O'Brien, S. A. (2019). Why Uber and Lyft drivers are striking. *CNN Business*.

Omazić, M. A., Labaš, D., & Uroić, P. (2023). Access Economy (Ethics). In S. Idowu, R. Schmidpeter, N. Capaldi, L. Zu, M. Del Baldo, & R. Abreu (Eds.), *Encyclopedia of sustainable management*. Springer.

Osman, H., D'Acunto, D., & Johns, N. (2019). Home and away: Why do consumers shy away from reporting negative experiences in the peer-to-peer realms? *Psychology and Marketing, 36*(12), 1162–1175.

Owyang, J., Tran, C., & Silva, C. (2013). *The collaborative economy*. Altimeter.

Palgan, Y. V., Zvolska, L., & Mont, O. (2017). Sustainability framings of accommodation sharing. *Environmental Innovation and Societal Transitions, 23*, 70–83.

Pan, Y., & Qiu, L. (2022). How ride-sharing is shaping public transit system: A counterfactual estimator approach. *Production and Operations Management, 31*(3), 906–927.

Pargman, D., Eriksson, E., & Friday, A. (2016, June). Limits to the sharing economy. In *Proceedings of the second workshop on computing within limits* (pp. 1–7).

Parguel, B., Lunardo, R., & Benoit-Moreau, F. (2017). Sustainability of the sharing economy in question: When second-hand peer-to-peer platforms stimulate indulgent consumption. *Technological Forecasting and Social Change, 125*, 48–57.
Peltoniemi, M., & Vuori, E. (2004). Business ecosystem as the new approach to complex adaptive business environments. In *Proceedings of eBusiness research forum* (Vol. 2, No. 22, pp. 267–281).
Pignot, E. (2023). Who is pulling the strings in the platform economy? Accounting for the dark and unexpected sides of algorithmic control. *Organization, 30*(1), 140–167.
Pouri, M. J., & Hilty, L. M. (2021). The digital sharing economy: A confluence of technical and social sharing. *Environmental Innovation and Societal Transitions, 38*, 127–139.
Quattrone, G., Greatorex, A., Quercia, D., Capra, L., & Musolesi, M. (2018). Analyzing and predicting the spatial penetration of Airbnb in US cities. *EPJ Data Science, 7*(1), 31.
Rabiei-Dastjerdi, H., McArdle, G., & Hynes, W. (2022). Which came first, the gentrification or the Airbnb? Identifying spatial patterns of neighbourhood change using Airbnb data. *Habitat International, 125*, 102582.
Rahim, N., Lepanjuuri, K., Day, F., Piggott, H., Hudson, R., & Lubian, K. (2017). *Research on the sharing economy*. London.
Rana, N. P., Kar, A. K., Gupta, M., Pappas, I. O., & Papadopoulos, T. (2023). Unravelling the dark side of sharing economy–Managing and sustaining B2B relationships on digital platforms. *Industrial Marketing Management, 113*, A4–A10.
Ranjbari, M., Morales-Alonso, G., & Carrasco-Gallego, R. (2018). Conceptualizing the sharing economy through presenting a comprehensive framework. *Sustainability, 10*(7), 2336.
Rinne. (2017). *What exactly is the sharing economy?* Retrieved from: https://www.weforum.org/agenda/2017/12/when-is-sharing-not-really-sharing/
Rojanakit, P., de Oliveira, R. T., & Dulleck, U. (2022). The sharing economy: A critical review and research agenda. *Journal of Business Research, 139*, 1317–1334.
Roma, P., Panniello, U., & Nigro, G. L. (2019). Sharing economy and incumbents' pricing strategy: The impact of Airbnb on the hospitality industry. *International Journal of Production Economics, 214*, 17–29.
Roover, H. (2016). *A framework to analyse a sharing economy*. Wageningen University and Research Centre.
Rothschild, M. (1990). *Bionomics: Economy as ecosystem*. Henry Holt.
Schaefers, T., Wittkowski, K., & Benoit (nee Moeller), S., & Ferraro, R. (2016). Contagious effects of customer misbehavior in access-based services. *Journal of Service Research, 19*(1), 3–21.
Schlagwein, D., Schoder, D., & Spindeldreher, K. (2020). Consolidated, systemic conceptualization, and definition of the "sharing economy". *Journal of the Association for Information Science and Technology, 71*(7), 817–838.
Schor, J. (2016). Debating the sharing economy. *Journal of Self-Governance and Management Economics, 4*(3), 7–22.
Si, S., Chen, H., Liu, W., & Yan, Y. (2021). Disruptive innovation, business model and sharing economy: The bike-sharing cases in China. *Management Decision, 59*(11), 2674–2692.
Sumter, D., Bakker, C., & Balkenende, R. (2018). The role of product design in creating circular business models: A case study on the lease and refurbishment of baby strollers. *Sustainability, 10*(7).
Sutherland, W., & Jarrahi, M. H. (2018). The sharing economy and digital platforms: A review and research agenda. *International Journal of Information Management, 43*, 328–341.
UN. (2020). *Frontier technology quarterly: Does the sharing economy share or concentrate?* Retrieved from https://www.un.org/development/desa/dpad/publication/frontier-technology-quarterly-does-the-sharing-economy-share-or-concentrate/
Vith, S., Oberg, A., Höllerer, M. A., & Meyer, R. E. (2019). Envisioning the 'sharing city': Governance strategies for the sharing economy. *Journal of Business Ethics, 159*(4), 1023–1046.

Von Hoffen, M., Matzner, M., & Chasin, F. (2015, July). Designing an ontology-based web directory for the discovery of sharing and collaborative consumption platforms. In 2015 *IEEE 17th conference on business informatics* (Vol. 1, pp. 108–115).

Wang, Y., Xiang, D., Yang, Z., & Ma, S. S. (2019). Unraveling customer sustainable consumption behaviors in sharing economy: A socio-economic approach based on social exchange theory. *Journal of Cleaner Production, 208*, 869–879.

Wilhelms, M. P., Merfeld, K., & Henkel, S. (2017). Yours, mine, and ours: A user-centric analysis of opportunities and challenges in peer-to-peer asset sharing. *Business Horizons, 60*(6), 771–781.

Wirtz, J., So, K. K. F., Mody, M. A., Liu, S. Q., & Chun, H. H. (2019). Platforms in the peer-to-peer sharing economy. *Journal of Service Management, 30*(4), 452–483.

Wu, S., & Benson-Rea, M. (2024). Uncovering the dark side of the sharing economy from a provider's perspective: a bibliometric systematic review. *Journal of Business and Industrial Marketing, 39*(6), 1226–1243.

Yang, Y., Tan, K. P. S., & Li, X. R. (2019). Antecedents and consequences of home-sharing stays: Evidence from a nationwide household tourism survey. *Tourism Management, 70*, 15–28.

Zeng, S., & He, Y. (2023). Blessing or curse? Sharing economy and its impact on the community of customers and suppliers. *Decision Sciences, 54*(5), 514–534.

Zervas, G., Proserpio, D., & Byers, J. W. (2017). The rise of the sharing economy: Estimating the impact of Airbnb on the hotel industry. *Journal of Marketing Research, 54*(5), 687–705.

Zhou, L. (2024). *Why Lyft and Uber drivers did their largest strike ever*. Retrieved from: https://www.vox.com/politics/2024/2/14/24073382/lyft-uber-strike-valentines-day-2024

Chapter 3
To Share or Not to Share?

Approaching the Sharing Economy

3.1 The Drivers of Adoption

The richest subfield of studies on consumer behaviour in SE identifies the reasons why individuals turn to these platforms. This is because adoption plays a key role in determining SE success and scale-up while being central to a consideration of consumer reluctance (Khalek & Chakraborty, 2023). The achievement of a critical mass of users is pivotal in generating direct and indirect network effects on which SE platforms base a major part of their growth, and inadequate participation has often caused their failure. For this reason, understanding consumer motivation is of interest to both scholarly research and, to an even greater extent, managerial practice. The market scenario in which SE platforms operate has become populated over recent years, with large platforms steadily growing (e.g. Airbnb, Uber) and causing problems for smaller platforms and newcomers. Platform managers therefore need to gain knowledge about their actual and potential users to attract and retain them (Hawlitschek et al., 2018). Questioning why users adopt SE opens up a wide range of answers; the reasons might differ according to the type of shared assets, user characteristics, and the country where exchange takes place. The drivers of adoption for SE platforms can thus be analysed from different perspectives (Fig. 3.1).

C. Grieco, *Consumer Behavior in the Sharing Economy*, Contributions to Management Science, https://doi.org/10.1007/978-3-031-76279-6_3

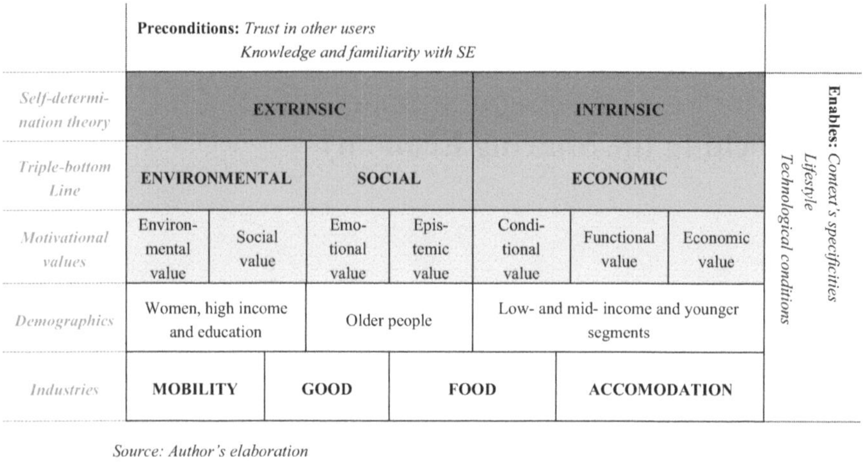

Fig. 3.1 A classification of sharing economy drivers. Source: Author's elaboration

3.1.1 Self-determination Theory

Studies focusing on SE motivations often turn to Deci and Ryan's (2000) self-determination theory (SDT) which strengthens the link between behaviour and motivation. According to this theory, individuals act from a need. Needs are the engine of actions as they incorporate the imbalance between current and desired states. Based on SDT, there are three fundamental needs potentially motivating individuals to carry out determined actions: autonomy, affiliation, and competence (Deci & Ryan, 1985). The first refers to the desire to be the master of one's own decisions; the second to feel connected and interact with the surroundings; and the third to the search for effectiveness. These needs may be satisfied by two categories of postulated motivators: intrinsic and extrinsic. The former emerges from the inherent value or enjoyment that an individual derives from an activity; the latter refers to the external pressures the individual perceives, or to outcomes that are separate from behaviour (Hamari et al., 2016). Bellotti et al. (2015) describe SDT as a spectrum extending from a lack of motivation, through increasing identification with external values, to the pure delight derived from carrying out an activity. Lindenberg (2001) further details intrinsic motivations as both enjoyment of an activity and the value derived from acting according to norms. An intermediate type of motivation that emerges from SDT is termed introjected motivation and refers to the effect of cultural and social norms on the behaviour of the individual, that is, awareness of their importance, while these remain external to the individual's self-conception. In this sense, introjected motivations lie between intrinsic and extrinsic motivations, as their locus of causality is external and only partially internalized, yet they are often rooted in a feeling of responsibility and the desire to behave in conformity with social expectations and norms (Deci et al., 1999). Examples of introjected motivations are societal and religious doctrines (Hellwig et al., 2015).

The appropriateness of adopting this theoretical framework in the context of SE lies in the type of exchange activities occurring in sharing initiatives that meet the fundamental needs proposed by Deci and Ryan (1985) and Alzamora-Ruiz et al. (2020). The need for autonomy is fulfilled by the innovative and more effective way of consuming offered by SE. Moreover, individuals may experience the gratification that comes from the autonomy and freedom offered by these platforms. The need for affiliation is met through social interactions between SE users and the creation of a community that often shares values and interests. Finally, accessing economically beneficial goods and services allows individuals to satisfy the need for competence. This is true also from the providers' perspective, as SE promotes additional revenue by sharing owned assets whose capacity would otherwise be wasted. As to the different motivators, intrinsic motivations have a stronger impact on attitude formation but have less influence on actual behaviour. The latter is a more significant predictor of continuous use intention. For this reason, there might be the risk of a crowding-out effect, where the intrinsic motivations that initially drive individuals towards SE are thwarted by extrinsic motivations that become stronger the more the platforms are used (Hamari et al., 2016).

3.1.2 The Triple Bottom Line

Consistent with the motivations of SDT, Böcker and Meelen (2017) further classify SE drivers based on the dimensions of the triple bottom line, considering economic reasons to be the main extrinsic motivation, and social and environmental reasons to be intrinsic. The opportunity to save money is seen as the key driver in pushing potential and actual consumers towards SE, as access fees are usually lower than acquisition costs (Bardhi & Eckhardt, 2012; Möhlmann, 2015). This is true not only for consumers but also for providers who can increase their revenues with the monetary compensation from sharing the unused capacity of their asset. Moreover, these earnings allow providers to reduce the burdens of financial ownership by distributing fixed and associated costs (Zhang et al., 2019). Environmental reasons appear to be more controversial, with some studies suggesting that sensibility towards a sustainable lifestyle is a priority for sharing users (Piscicelli et al., 2015; Khan et al., 2020; Serrano et al., 2021), while others frame sustainability as a positive side effect rather than a main argument (Hartl et al., 2018; Guglielmetti Mugion et al., 2019). Starting out from this consideration, Merino-Saum et al. (2023) carried out an analysis of extant contributions on SE drivers to understand the role of sustainability-related reasons in orienting individuals towards these platforms. What they found is that environmental factors are not the predominant focus of most previous studies; rather, they were secondary to functional, social, hedonic, and economic factors. Their research also shed light on the meaning of environmental drivers, identifying four subfactors: expected environmental impact, biospheric/anti-environmental values, past green behaviour, and green practices conducted by the platform. Among them, the expected environmental impact is the factor most frequently mentioned, and in some cases, the typology of impact is further specified (e.g. consumption of natural resources, waste generation, carbon emission, water use) (Merino-Saum et al., 2023).

Social reasons refer to the interactions between consumers and providers that take place through the platform and differentiate SE from traditional consumption practices. These include different types of interactions such as meeting new people and communication and collaboration between users. The interest in community belonging and expansion of social connections has been a powerful driver towards SE, even more so in the first years of its development; however, as market mediation has increased, it has weakened compared to the opportunistic utilitarian motive (Habibi et al., 2017). In fact, the facilitation of digital technologies allows for the ubiquitous availability of assets through the platform, independent of user location and freeing parties from the need to be physically present at the same time, thereby weakening the role of social interaction in the transactions (Neunhoeffer & Teubner, 2018).

3.1.3 Motivational Values

Going beyond the dimensions of the triple bottom line, there are further specific values that can be considered as impacting on individual decisions. Khalek and Chakraborty (2023) identify four additional factors, beyond economic, social, and environmental values, which promote SE adoption from the consumer perspective: functional value refers to the perceived usefulness and relative advantage and ease of use of SE; emotional value focuses on the enjoyment and fun of using SE, also through wider product choice and experientiality; epistemic value depends on the novelty of SE that meets the need to broaden cultural boundaries; and conditional value sums up the context, personal situation, and environmental circumstances in which the individual approaches SE. The value of enjoyment is further confirmed by Hamari et al. (2016). The authors rely on studies on information sharing on the Internet (e.g. social networking), where the pleasure of participation emerges as a powerful initial motivation without predicting continued use. By investigating enjoyment in the context of SE, they find that it plays a key role in both attitude formation and in use intention. This shows that consumers turn to SE because it is fun and provides an important way to interact with others (Hamari et al., 2016). Variety-seeking is a further aspect related to enjoyment: individuals turn to SE for the abundant selection of products they can access; this is particularly the case with second-hand markets (Kim & Jin, 2020). These hedonic dimensions (i.e. enjoyment, fun, variety-seeking) are, surprisingly, closely associated with the economic motivation. Cost saving can add further enjoyment to the sharing process and provide consumers with emotional rewards for their smart shopping (Jin et al., 2003). Functional value is closely linked to the convenience of using SE. This is particularly relevant in contexts where the access-over-ownership principle behind SE platforms obviates the burden of possessing an asset with its attendant responsibility and costs (Dall Pizzol et al., 2017). Yet, this does not apply to all categories of shareable assets: while car sharing can be more convenient, especially in large cities, renting clothing is not perceived as a better option compared to shopping for new clothes (Kim & Jin, 2020).

3.1.4 Sectoral Differences

What is often specified in studies focused on SE drivers are sectoral differences. In fact, the different types of motivations associated with SE always concentrate on a specific industry. This is because of the variety of shareable assets that differ greatly in terms of economic value, expected environmental impact, or degree of social interactions with other users, and thus display different motives for being chosen. According to Böcker and Meelen (2017), a key element lies in the costs associated with the ownership of the asset or compared to the non-sharing alternative (e.g. traditional hotels): the higher the price, the greater will be the impact of economic motives in driving consumers. For this reason, house and car sharing are often chosen for the possibility of saving money while being rarely associated with social motives and almost never with environmental concerns (Merino-Saum et al., 2023). Environmental aspects emerge as more important in the mobility sector and in the recirculation of goods. As to the former, this largely depends on the well-known negative impact that car production has on the environment and on shared knowledge of the potential of car sharing to alleviate this impact. Nevertheless, this relationship is not always confirmed, as some studies on car sharing point to the predominance of utilitarian factors in orienting consumers' approach (Bardhi & Eckhardt, 2012; Guglielmetti Mugion et al., 2019). Social motivations have been investigated more often in goods and food sharing, where the SE form itself clearly involves more social interaction. Specific consideration relates to second-hand acquisitions, as the wide interpretation of SE covers the redistribution of goods. Guiot and Roux (2010) identify economic and recreational reasons as their main motivations, particularly in lower-income urban neighbourhoods characterized by significant budget constraints for individuals. On the other hand, second-hand purchasing also offers social benefits through the high level of interaction between providers and end consumers, as well as the pleasure gained from hunting for bargains, finding unique goods, and having access to a wider variety of choice compared to traditional channels (Guiot & Roux, 2010).

3.1.5 Demographic Differences

Demographics have also been considered in studies on SE drivers. According to the findings of Khalek and Chakraborty (2023)'s review, young educated urban males are typical SE adopters, while according to Hellwig et al. (2015) women show a greater disposition to share, with a higher number of shared objects and frequency of sharing. Consistent with this, Neunhoeffer and Teubner (2018) point to the lack of importance women place on ownership-related aspects which may drive their higher use of SE services. In general, traits such as intelligence, openness to experience, extraversion, and cosmopolitanism characterize SE consumers (Tunçel & Özkan Tektaş, 2020). Böcker and Meelen (2017) reveal how low- and mid-income segments are attracted more by the possibility of saving money and accessing assets otherwise out of their reach. Economic motives also appeal to younger consumers who are, at the same time, less socially motivated, as well as higher income and higher educated groups. Moreover, younger segments of the population are more prone to adopt SE because of their greater familiarity with innovative technologies. The younger generation exhibits a greater acceptance of new services in general and feels confident about their ability to use new technologies (Kurisu et al., 2021). Young individuals are also more accustomed to SE interaction procedures that differ from the traditional consumption style, while the older generation may find them more difficult to accept and use (Kurisu et al., 2021). On the other hand, older people are attracted more by social motives, although belonging to the group that shares least (Hellwig et al., 2015). Focusing on generational differences, Fan et al. (2023) reveal how consumer motivations towards peer-to-peer accommodation differ among the cohorts: individuals from Generation Z are influenced more by sustainability, social interaction, and social media exposure, while Millennials' need for self-improvement impacts on their evaluation of the service. Environmental reasons are powerful motivators for women. However, the authors stated that, compared to other differentiating criteria (e.g. sectorial differences), demographic traits have a lower impact on determining individuals' motives in sharing.

3.1.6 Preconditions and Enablers

Besides these drivers, scholars also identified a set of elements that contribute to SE adoption, regardless of the sectoral or demographic specificities of individuals. For example, according to Hawlitschek et al. (2016), trust in other users is a key precondition of SE usage (see par. 2.2), while Neunhoeffer and Teubner (2018) stress the role of knowledge and familiarity with SE platforms. Technological conditions also play a role in supporting SE adoption (v and Teubner, 2017) as well as individuals' lifestyle and the specificities of the context (Hellwig et al., 2015). These elements appear to be closely interrelated. Extended participation in SE leads to increased familiarity with the platform that, in turn, implies stronger engagement with it. This decreases both the impact of trust issues and the perception of the expected effort required with technology use that might otherwise deter use of these platforms. On the other hand, rare use implies that coordination and search costs can exceed the perceived utility of SE platforms and hold back user segments from their adoption (Lamberton & Rose, 2012; Möhlmann, 2015). Digital access is, however, a key prerequisite and, regardless of the perception of effort, is objectively linked to geographical context. It is worth noting that a large number of contributions on SE drivers focus on adoption within specific countries and regions (Table 3.1). This is

Table 3.1 Drivers of sharing economy adoption within specific geographical contexts

Author (Year)	Industry	Country	Drivers
Lee et al. (2021)	Mobility	USA–China	Price, relational values USA: quality, perceived regulation China: previous technological experience
Chatterjee et al. (2019)	Accommodation	India	Price, facilities, home-like feeling, trust, friendly service, perceived authenticity, social norms, and effort expectancy
Wu et al. (2017)	Accommodation	China	Utilitarian motivation, hedonic motivation, and perceived trust do have positive effects on tourists' behavioural intentions. Past experience with room-sharing moderates these effects
Tsai et al. (2021)	Mobility	Bangkok	Sustainability, enjoyment, economic benefit
Ek Styvén and Mariani (2020)	Second-hand	UK	Perceived sustainability, economic motivations, and distancing oneself from the consumption system
Xu et al. (2021)	Accommodation	USA	Information disclosure: namely, what information (i.e. information content), from where (i.e. information source), in what format (information presentation format), and how much (the quantity of information)
Violin (2021)	Mobility	Brazil	Technological availability, expectation of benefits, and the reduced fee
Burinskienė et al. (2021)	General	USA	Wage level, social network users, import level, and personal consumption
Rebiazina et al. (2018)	General	Russia	Being young (up to 35 years) and living in large Russian cities; interest, comfort, utility; approval by the reference group (family, friends); ecological and environmental benefits; ease of use; high level of trust, environmental protection, cost reduction, additional income, modern lifestyle, and simplicity of use
Hinojo et al. (2022)	Second-hand	Spain	Being male, relatively young, with children, a frequent Internet user, in employment, and living in a household with some price-consciousness and environmental awareness. The scarcity of bricks-and-mortar stores
Jain and Mishra (2020)	Luxury	India	Economic benefits, fashion involvement, self-pleasing experience, social projection, perceived risk, and past sustainable behaviour
Aguilera-García et al. (2021)	Mobility	Spain	Age, occupation, income, and environmental awareness
Albinsson et al. (2019)	Cross-sector	USA and India	Perceived sustainability is the strongest predictor of usage followed by trust, generosity, risk seeking, materialism, power distance, long-term orientation, and collectivism (no differences among respondents from the two countries)

(continued)

Table 1 (continued)

Author (Year)	Industry	Country	Drivers
Kurisu et al. (2021)	Cross-sector	Japan and Switzerland	Prior usage experience, younger segments with higher materialism and normative tendencies are more likely to adopt the services. The maturity of the service in the country determines usage and motivation
Borusiak et al. (2021)	Second-hand	Spain and Poland	Environmental concerns influence attitudes (more in Spain than in Poland), but the ascription of responsibility has a low impact in both countries. The awareness of the consequences has a high impact on attitude in both the countries
Grybaitė and Stankevičienė (2016)	General	Lithuania	Making extra money, the possibility to support individuals and/or companies; meeting new people and having an interesting experience. The survey reveals a shared preference towards owning things rather than sharing them among respondents

Source: Author's elaboration

Insights: Drivers of Adoption in Collaborative Fashion for Generation Z – Second-Hand Shopping Versus Rental Commerce

The following figures represent the result of an empirical investigation into the drivers of adoption of collaborative fashion platforms. A questionnaire was distributed through Google Form and self-administered by a sample of 387 individuals from Generation Z. The drivers investigated are derived from existing literature on SE motives for adoption, namely, sustainable lifestyle, economic benefits, anti-consumption, and social interactions. Some specific elements have also been added based on research into purchasing drivers in the fashion industry, (i.e. the search for uniqueness and the quality of the products) and for the specific target addressed (consumers from Generation Z). SE in the fashion industry includes different business models. In this research, the objects under investigation are rental commerce (access-over-ownership) and second-hand shopping (transfer of ownership).

The findings reveal that respondents exhibit a similar approach to both services in terms of purchasing drivers. Sustainability concerns play the greater role in orienting Generation Z consumers towards these SE services, consistent with the label of "greener generation" ascribed to them. This seems to be slightly more important with second-hand shopping compared to rental commerce.

About two-thirds of the sample agree on the importance of saving natural resources, adopting a sustainable way of consuming, and respecting the environment as key factors to choose these services.

Sustainability

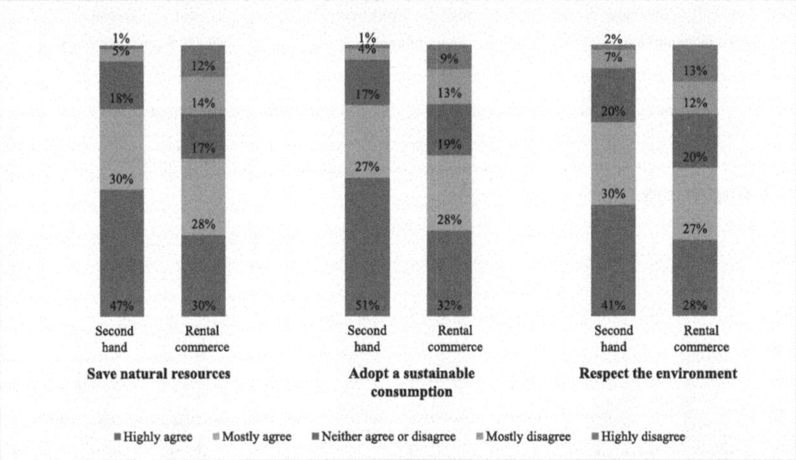

The economic driver points to two interesting considerations. Firstly, respondents state that they see these services as a way of affording more garments, as their prices are lower than in traditional shopping. This is mostly true for second-hand shopping, but it is noticeable also in rental commerce.

Economic benefits

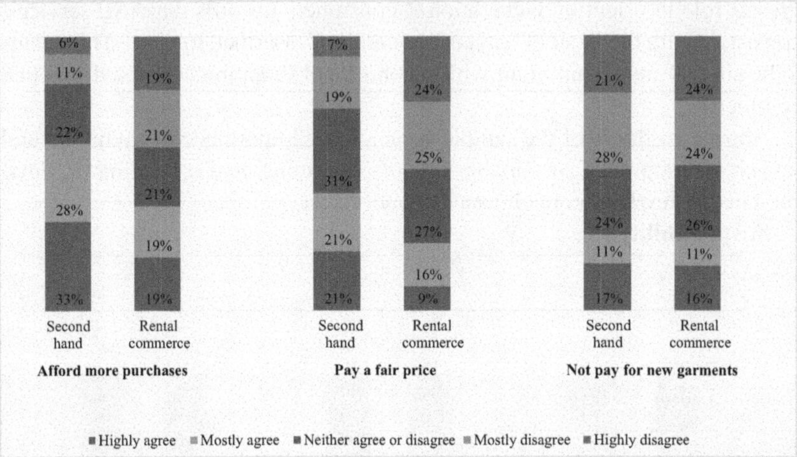

Uniqueness

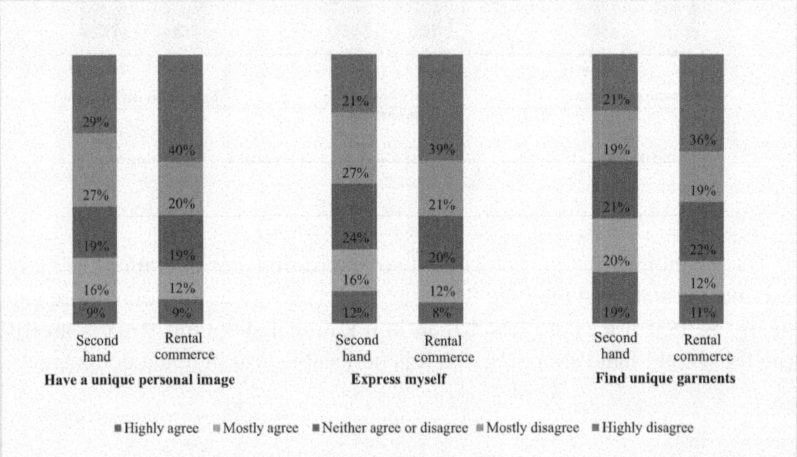

This gives an interesting angle on the recently raised issue of the link between SE and indulgent consumption, according to which some aspects of SE stimulate behaviour that contradicts its premises. This is particularly evident in the example of second-hand shopping, where affordable prices may lead consumers to buy more than necessary and eventually lead to more purchases instead of reducing consumption. A further consideration concerns the possibility of paying a fairer price for garments, again more prominent for second-hand shopping while less significant in rental commerce.

A further driver that characterizes the approach to these platforms is the possibility of having access to unique or particular garments. This emerges as important for second-hand shopping where vintage items can be found. Conversely, renting clothes is seen as a way of following trends by wearing fashionable clothes without owning them. In this respect, the goal is not to be unique but to have convenient access to trendy, often luxury garments.

Social relations

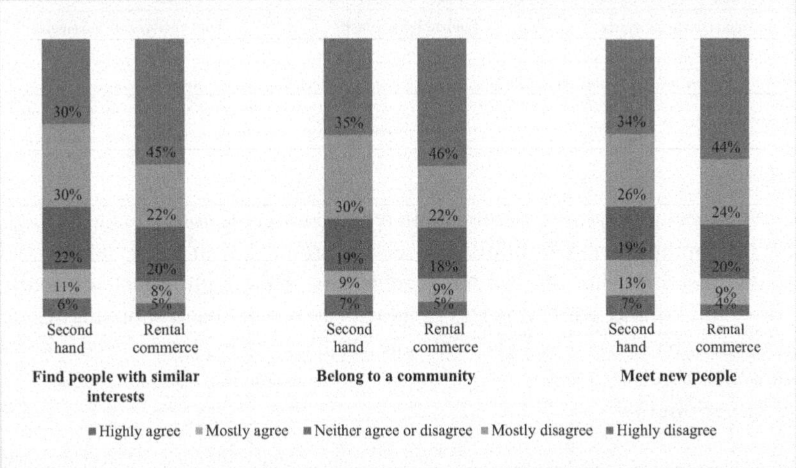

The search for social interaction is the least important aspect driving respondents to both second-hand shopping and rental commerce. While in other industries, such as shared mobility or accommodation, it can play an important role, these services are usually highly mediated by the platform, making exchanges almost totally devoid of face-to-face encounters. The platforms usually allow for direct contact between providers and consumers (e.g. chats and messages), yet this is not a primary concern for those using these services.

Quality

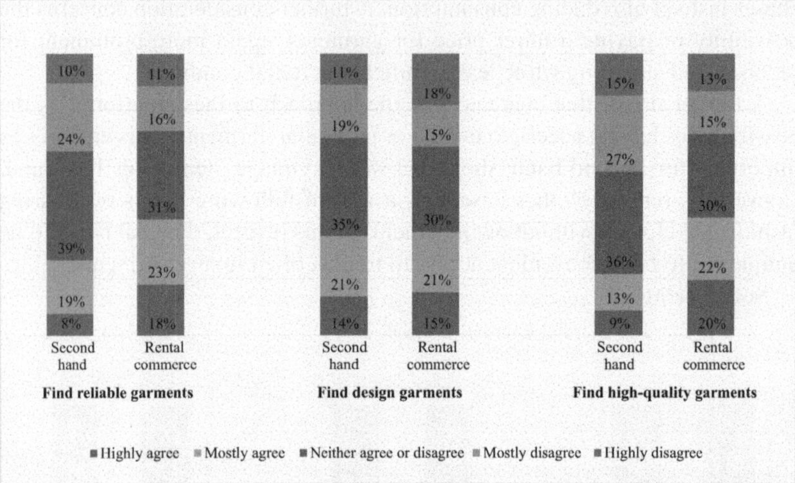

An aspect that is more significant in rental commerce compared to second-hand shopping is the possibility of accessing products with higher quality and design, also ensuring the garments' reliability. This is important for nearly half of the sample, while it is less of a factor in second-hand shopping where consumers buy pre-owned garments and are likely to expect reduced quality for a low price.

Functionality

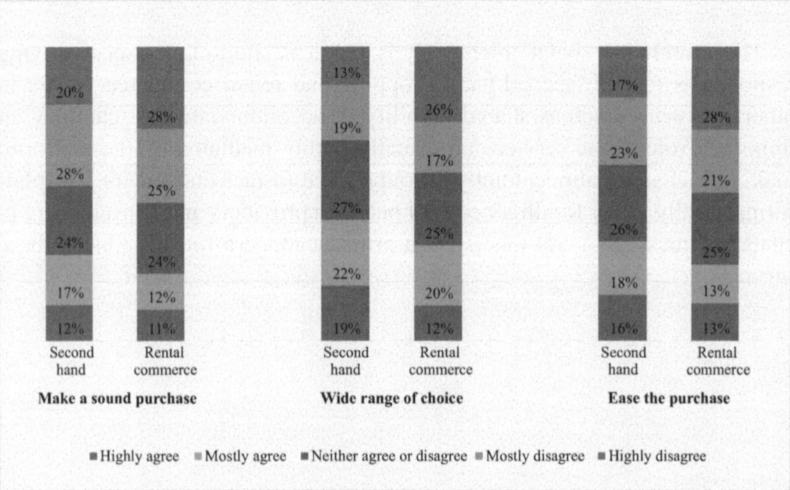

The last driver refers to the role of the digital platform and the possibility of using these services through online channels rather than by traditional retail. Overall, it appears that this aspect does not have a strong impact on the purchasing decision. This may depend on the specific target of this research, i.e. Generation Z, who are very much used to online shopping, such that it has no specific value in orienting their choices.

because legislation, cultural values, and contextual habits shape the conditions in which SE develops, as well as the characteristics of its traditional competitors and potential and actual consumers.

3.2 The Trust Issue

A key topic in SE as a form of online interaction is that of trust. According to Mayer et al. (1995), trust is the confidence that the other party will not exploit one's vulnerability; it depends on the ability of the trustees to keep promises made, the benevolence of their intentions, and the integrity of their behaviour. Despite also being essential to traditional B2C transactions, what makes trust a delicate topic for SE is the fact that exchanges take place between users who do not know each other, and for this reason, it can become the main barrier to user participation (Schor, 2016). Moreover, SE is characterized by high levels of risk, uncertainty, and interdependence, as a result of which trust assumes increased importance (Ter Huurne et al., 2017). On the other hand, it might be easier for users to establish trust with unknown people as, in the event of failure, there is no risk in ruining existing relations. This is what leads to the building of strong relationships with others and the engagement in mutual aid behaviours (Lang et al., 2020).

The reasons why trust is such a delicate area within SE lie in the specificity of its business model, where both providers and consumers are mostly private individuals whose relationship takes place through an online platform. Peer-to-peer exchanges imply the creation of economic and social bonding between users. In SE, consumer (but also provider) protection is occasional, with rules and practices mostly left to self-regulation and with an unclear division of responsibility between platforms and providers about consumer disputes (Berg et al., 2020). For this reason, it is essential that platforms adopt appropriate mechanisms to increase trust and, in turn, reduce potential hazards (Ponte et al., 2015). Marketing research into offline purchases reveals the extent of the impact first impressions of a salesperson have on trust formation and the perceived trustworthiness of the firm (Wood et al., 2008), which in turn increases purchase intention and customer loyalty (Kim et al., 2008; Chen & Quester, 2015). Online exchanges cannot benefit from these direct interactions between sellers and buyers and are thus required to adopt and implement appropriate strategies to build and nurture trust among users (Marth et al., 2022). Moreover, for-profit platforms can rely on reputational mechanisms and external warranties to gain users' trust. In this context, trust does not solely depend on a decision process;

rather, it can be triggered by elements such as the brand (Castelfranchi & Falcone, 2010) or the belief in application of laws and regulations by public authorities (Berg et al., 2020). Non-profit platforms can instead leverage users' intentions to create a common philosophy or mindset and a sense of belonging that supports relations and collaboration.

It is worth noting that while, traditionally, trust refers to a dyadic relationship between trustor and trustee, in SE this concept has a different configuration with three key actors involved: the supplier, the demander, and the platform. As a result, there are more levels of trust that need to be ensured. Möhlmann (2015), for example, identifies two different levels: interpersonal trust and institutional trust. The former in SE implies the openness of a person to unknown people. Interpersonal trust refers to peers' relations within the platform, whose collaborations and exchanges create the final value. For this reason, this type of trust is full of expectations of both trustor and trustee (Simpson, 2007). Sharing platforms entails a level of relationship that is greater than, for example, e-commerce. While the latter is usually limited to communication or delivery, using SE platforms can imply travelling or eating with other users, or staying at someone's house. A further typology of trust is the institutional one, which in SE plays an equally important role. Institutional trust is the perception of reliability and credibility that users develop towards the platform, which acts as the firm behind the product/service exchange. Choudhury and Karahanna (2008) distinguish between two aspects of institutional trust. Informative trust is made up of the beliefs and convictions of users about the truthfulness and accuracy of provided information. Structural trust refers to the perception users have about the reliability of the technology in terms of safety and privacy concerns. Both these elements are crucial in the first purchase where familiarity is still low. Institutional trust becomes essential in facilitating transactions on digital platforms, and it can be improved with external certifications and further warranties. According to Möhlmann (2015), there is a spill-over effect by which trust in the platform provider leads to trust in the users participating in the exchanges on that platform in a hierarchical way. In fact, following the trust transfer theory, trust can pass between different targets that are perceived as associated. This emerges as being particularly important where there are low levels of trust, such as at the beginning of the relationship. When consumers perceive a platform as trustworthy for example because of its large network size, or the review system it is based on, they will be more likely to perceive the same trust towards the providers of that platform. However, trust can also be lost according to the same principle (Grüner et al., 2024).

While confirming the importance of institutional-based trust, Hofmann et al. (2017) focus on reason-based trust which, from their perspective, is overlooked in the literature on SE. Reason-based trust includes calculative trust, rational trust, and knowledge-based trust (Castelfranchi & Falcone, 2010). In the SE context, it implies that a perception of the platform as rational gives consumers reasons to trust and consequently adopt it (Kansal & Bhalla, 2022). Reason-based trust develops over time based on consumers' assessments of their experiences. When individuals realize that the other party pursues similar objectives, acts benevolently, has the appropriate competences, and is supported by an external third party, they are inclined to trust it. In this sense, reason-based trust differs from implicit trust, that is, an automatic reaction to specific triggers and to good experiences (Hofmann et al., 2017). Botsman (2017) explains trust in SE introducing the concept of distributed trust, which has been made possible through digital technologies. In her view, three phases characterized the evolution of trust in human relations. The first is the local trust that was typical of small communities where everybody knows each other. The second is the institutional trust that is intermediated by the societal institutions of the industrial societies. The distributed trust on which SE is based leads strangers to trust each other because of their connection through digital technologies, with reputation being the main currency (Berg et al., 2020).

Trust-building measures can be developed to signal trustworthiness to potential consumers. SE platforms, in their role of trustee, should continually introduce new and innovative ways to support the creation and nurturing of trust. This is also because trust mechanisms have a cumulative effect: the more platforms work on it, the greater is the possibility that trusting relationships develop (Möhlmann & Geissinger, 2018). In the context of online transactions, these measures include information about payment security, direct support channels, a reputation system such as graphic ratings and reviews, and strict regulations and sanctions for those not behaving correctly (Marth et al., 2022). In the SE domain, platforms have gradually adopted these measures, particularly reputation systems that assess both suppliers and demanders, to help users deciding on their adoption and, at the same time, limiting the participation of unskilled or dishonest people (Resnick et al., 2000). As for regulations, the critical element is that SE largely relies on self-regulation practices, with the platform establishing precise behavioural standards for both providers and consumers that are often left to the control individuals exert on themselves and other users. Kong et al. (2020) identify three key mechanisms of trust in SE which, in turn, can increase user continuance intention: (i) social referrals, that is, when consumers share their purchase experience to offer a perspective to other potential consumers; (ii) information quality, or the extent to which disclosed information is updated, accurate, useful, and complete; and (iii) transaction safety, namely, the idea that safety and privacy are always ensured and any risk is minimal. Marth et al. (2022) carried out an analysis of SE platforms in the accommodation industry to assess the presence of trust-building measures on their websites and the extent to which these can actually have an impact on consumers' perception of platform trustworthiness. What they came up with is that, in the platforms analysed, there are a small number of community features (i.e. tools to interact with other users) and poor use of external quality seals regarding platforms' trustworthiness or technical flawlessness. On the other hand, all the platforms use reputational systems allowing consumers to rate their experience and write comments. At the same time, an analysis of the platforms' terms and conditions reveals minimum regulatory control: platforms fail to regularly perform key tasks such as monitoring identities and contents. Interestingly, their research on consumers shows that, while regulations play a key role in increasing willingness to engage with the platform, trust-building measures on the website do not have a significant impact on reducing consumers' perceptions of risks and are thus ineffective in supporting participation. This is probably due to the association between trust and reputation which leads platforms to focus on the use of reputational systems alone. Yet, despite being pivotal, reputation does not reflect the complexity of the trust concept which extends beyond it (Ter Huurne et al., 2017).

Case Study: Vinted and Its Trust Flows[1]

The rise of Vinted follows the significant growth of apps that allow buyers and sellers to exchange second-hand items, especially clothes and fashion accessories. The company was founded in 2008 in Vilnius, Lithuania, from the concept of two young local entrepreneurs, Milda Mitkute and Justas Janauskas, who needed to clear out their wardrobe before moving. They therefore created a swapping platform to help declutter surplus clothes. Today, Vinted has more than 5000 employees, more than 300 million products available, and more than 50 million registered users who sell, buy, and trade their second-hand clothes. It is currently one of the leading companies in the second-hand goods market and competes with established fashion brands.

The importance of trust for SE in general increases when second-hand goods are concerned: users must be further ensured about the condition of the goods. Vinted defines itself as a community based on trust, honesty, and friendliness; analysis of the platform allows us to identify the four main flows of trust, and the measures adopted to create and maintain them (Fig. 3.2).

1. **Platform-users trust.** The first trust flow occurs between the platform and its user base with the goal of establishing initial trust at a general level. To this end, Vinted invests in television advertising with the precise intention of generating trust in potential users: many people believe that the presence of a brand in television advertising spaces makes it more trustworthy than other brands. With the same purpose, influencer marketing campaigns have been developed by the platforms, gaining considerable success. A further element in this trust flow is that of transaction security, a fundamental requirement for the existence of any marketplace. A totally secure shopping experience for Vinted users is guaranteed by integrated payment and shipping services, which allow them to make purchases and conveniently handle shipping issues without having to leave the platform. The integrated system has been set up to safeguard privacy, so that personal information entered on the platform is not shared with other members unless they are needed for order processing. Security has also to be guaranteed at a personal level, creating policies and tools that promote community success for both sellers and buyers. If users violate the brand's

[1] Data have been collected through semi-structured interviews to Vinted marketing managers and a desk research on the Vinted website, press releases, and reports.

terms and conditions, use a fictitious identity or sell banned items, they can be warned, removed, or deactivated completely. The Vinted team uses software that detects private messages deemed inappropriate and immediately hides them from the user to whom the message was sent. It also uses predictive analytics to identify activities considered suspicious to block them (e.g. uploading of counterfeit luxury goods).

2. **Platform-providers trust.** From the providers' point of view, trust in other users plays a secondary role, as their main motivation is to make money in a quick and easy way. Consequently, sellers will choose the platform that has the highest number of users and is easiest to use. In this case, trust is first and foremost about payment systems, which have the responsibility of checking the identity card and the Iban code of the buyers. On the provider side, trust is connected to the quantity and rapidity of selling goods. The effort of Vinted is thus to guarantee that they will be repaid for the efforts made to offer a complete and attractive presentation of the items for sale, as their trust depends on their reward. The Vinted algorithm manages to give visibility even to users that have few followers; thus, there is no need for sellers to have a community behind them that can support their visibility, as is the case with other platforms (e.g. Depop). In its relations with sellers, an important value for Vinted is brand adherence: once the user becomes accustomed to using a certain platform, a sort of bond is created, and this bond leads to extended use. Vinted's main objective, then, becomes that of helping sellers upload and sell their goods quickly, so that they remain satisfied and continue to use the platform. The key is to turn the beginner into a seller in the shortest possible time: as soon as the newly signed-up sellers perceive the possibility of continuing to sell successfully on the platform, they can be considered retained. Vinted leaves a great deal of autonomy to the seller, who can choose a different courier and indicate the shipping price to the buyer, who is always responsible for the shipping costs.
3. **Platform-buyers trust.** A key role is played by the buyer protection fee charged on every transaction, enabling services such as customer support, the integrated secure payment system, and the refund policy for delivery problems or items not matching the description. The fixed fee is used to cover fixed costs such as processing payments or providing the shipping label and is 0.70 € per transaction. In addition, a variable commission of 5% is charged (if the item is sold for 100 €, the buyer has to pay an additional 5 €). In addition to being a source of revenue, this tool makes buyers feel protected, gives the platform an image of a guarantor, and, consequently, makes people more likely to use it. The amount paid by the buyers is not immediately transferred to the seller; rather, it is marked as "outstanding amount" until the transaction is completed. The purpose of this mechanism is to ensure the security of users' payments: until the buyer has

declared that the purchased good has been correctly received according to its description, the money will be retained by the platform, so that disputes can be handled, and the seller can be refunded. For instance, if the purchased item does not match the buyer's expectations, the money spent can be returned.

4. **Sellers-buyers trust.** Despite the platform acting as an intermediary, users also need to know that they can trust other users when carrying out online transactions. The role of the platform thus is also that of creating the proper mechanisms to ensure each user that there are no hazards in dealing with others. A first element is that of users' public profiles, which contain user data and reviews posted by peers. As with the majority of online platforms, Vinted supports trust building by providing review mechanisms through which members can rate other users based on their experiences. Regarding information quality, the rating mechanism on Vinted uses stars: people are advised to give ratings (from one to five stars) and to write comments every time a transaction is made in order to signal to other users the level of trustworthiness of a particular buyer/seller. Vinted works on educating their users, providing them with guidelines on how to become a trustworthy counterpart of the transaction. With this aim, there is a section on the Vinted website which is entitled "Let's help create trust in our community". The tips contained cover four different topics: communication, reviews, verified accounts, and initiative.

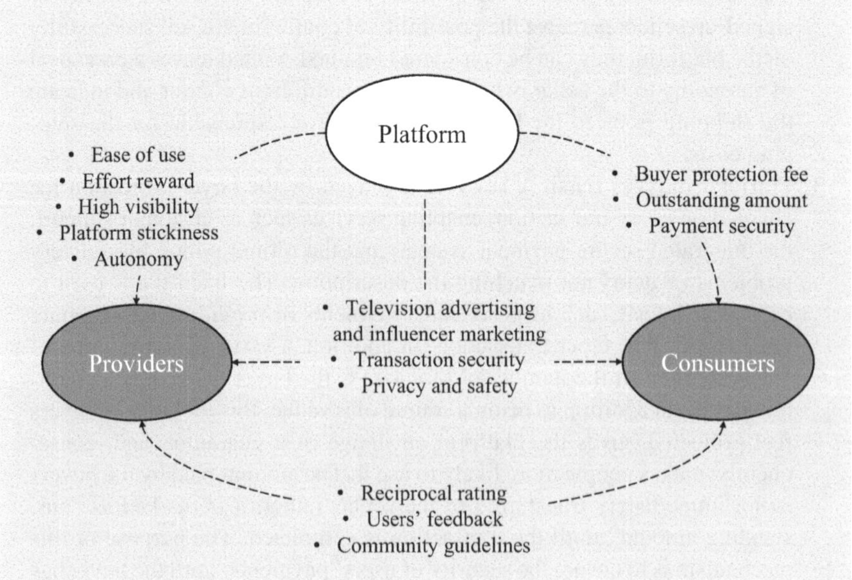

Fig. 3.2 The trust flow of Vinted. Source: Authors' elaboration

3.3 Examples of Customer Segmentation Schemes

The analysis of SE drivers sheds light on a heterogeneous scenario where motives different in nature (e.g. utilitarian versus prosocial) can shape the attitude of potential and actual consumers. Based on this heterogeneity, scholars have attempted to define the prototypical profiles of sharing consumers, using the principles of the segmentation process to cluster individuals according to their approaches to SE (Table 3.2). A first example comes from Hellwig et al. (2015) who carried out a mixed-method analysis with qualitative and quantitative investigations to identify the key variables to segment sharing consumers and group them into homogeneous

Table 3.2 The features of sharing economy consumers

Frequent usage	Average usage		Low or no usage
Sharing idealists [F, Pt]	**Sharing normatives**	**Sharing pragmatists** [M, Shs, Ft]	**Sharing opponents** [M, Ft]
Intrinsic motivations	Introjected motivations	Introjected motivations	Extrinsic motivations
+ Generosity General reciprocity – Perfectionism	+ Social media usage Perceived resource scarcity	– Generosity General reciprocity	+ Perfectionism Tit-for-tat reciprocity – Social media usage
Change seekers [Lhs, Hi, He]	**Conscious materialists** [Li, Le]	**Premium keepers**	**Fickle floaters** [O]
+ Seek variety – Possessiveness Materialism	+ Economic consciousness Environmental consciousness Possessiveness Materialism – Brand loyalty	+ Status consumption Brand loyalty – Economic consciousness	+ Economic consciousness – Status search Brand loyalty
Committed	**Pragmatist**	**Intermittent**	**Sceptical**
+ Anti-consumption Environmental consciousness Community belonging	+ Utilitarianism – Environmental consciousness	+ Convenience – Community belonging	+ Traditional consumption
Social enthusiasts [F, Lhs, Y]	**Sceptic ascetics** [Lhs]	**Conflicted materialists**	**Individualistic refuseniks** [M, Shs]
+ Risk seeking	– Ownership-related aspects Car owners + Risk seeking	+ Ownership-related aspects – Risk seeking	– Risk seeking

(continued)

Table 3.2 (continued)

Frequent usage	Average usage		Low or no usage
Diverse platform sharers	**Power platform sharers** [Y]	**Mobility-focused sharers** [O]	**Non-sharers**
+ Social identity Innovativeness Sustainable lifestyle Behavioural nudging	+ Social identity Innovativeness – Behavioural nudging	+ Ability to test new product – Innovativeness Behavioural nudging	– Innovativeness Social identity Behavioural nudging

Legend for demographic characteristics: *M* male, *F* female, *Y* young, *O* old, *Hi* high income, *Li* low income, *He* high education, *Le* low education, *Lhs* large household size, *Shs* small household size, *Fu* full-time worker, entrepreneur, or manager, *Pt* part-time worker or homemaker
Source: Author's elaboration

clusters. In their research, the authors found how sharing consumers differ in terms of behaviour and disposition. The former refers to the type of object that individuals are willing to share, based on a personal hierarchy of relevance and intimacy, and to the more or less close circle of people they would share with. Sharing disposition depends on three groups of variables. The first are trait-related variables and include generosity in giving, perfectionism related to the object to be shared, and reciprocity, either direct or generalized. The second group concerns motivational variables and refers to the SDT distinction between intrinsic, extrinsic, and introjected motivation. The last group relates to the socio-economic self-perception of individuals which sums up their perceived resource scarcity and therefore the salience of sharing. Based on these variables, Hellwig et al. (2015) analysed a sample of 1121 SE users in Germany, clustering them into four segments with different sharing behaviour and dispositions. The sharing idealists are mostly women, part-time workers, intrinsically driven to share, with the highest scores for generosity and the lowest for perfectionism and direct reciprocity. The sharing pragmatists are mostly men, full-time employed or trained, who see sharing as a code of conduct among friends and family members. The sharing normatives are highly active on social media and use SE pushed by introjected motivations that lead them to see it as something socially desirable. Finally, the sharing opponents are mostly men, often independent entrepreneurs or managers, not particularly active on social media, who are not interested in SE and make poor use of it.

The same method is used by Lawson et al. (2016), who focus on access-based consumption practices to understand consumer motivations and cluster a sample of 220 potential and actual users. Their qualitative investigation reveals how economic and environmental consciousness, status consumption, choice confusion, possessiveness, and materialism are the key factors driving consumers. Moreover, the possibility of using a product flexibly and of having access to a wide variety of assets plays a role in determining consumer attitudes and purchase intentions. Clustering consumers based on these variables led the authors to identify four groups. The fickle floaters, with the lowest attitude and intentions towards SE, are highly impacted by economic considerations and have least interest in social status and brand loyalty. The premium keepers, with a positive attitude to SE but the lowest level of purchase intention, are interested in variety-seeking and possessiveness and do not care about spending money if they find a satisfying product/service. The conscious materialists have the lowest incomes and are simultaneously searching for both economically and environmentally sustainable choices. Finally, the change seekers, with the highest attitude and purchase intentions towards SE, seek variety more than the other clusters and exhibit the lowest scores on possessiveness and materialism.

A further attempt comes from Neunhoeffer and Teubner (2018) who focused on peer-to-peer sharing and clustered a sample of 745 actual and potential users across different industries through 18 distinct motives and impediments. In particular, they processed these variables into five second-order constructs: concerns, benefits, product-specific aspects, social aspects, and ownership-related aspects. These four segments reveal the role of elements such as car ownership, engagement duration, household size, and risk propensity in shaping behaviour towards SE. The first cluster is labelled social enthusiasts and shows the highest levels of sharing activities across the different industries investigated. This cluster is the youngest in terms of age, yet it comprises individuals with the longest experience of SE adoption. The second is the conflicted materialists cluster and is made up of users with the highest score for ownership issues who adopt SE while remaining highly concerned about its drawbacks. The sceptic ascetics, on the other hand, exhibit lower scores in both concerns and benefits while making extensive use of SE platforms, most of all within the accommodation and mobility industries. The last cluster is that of individualistic refuseniks who make scant use of SE platforms. It is interesting to note the role of household size in differentiating between the segments. This variable is positively correlated with the perception of benefits and the social aspects of SE while implying less pronounced concerns and lower interest in ownership-related aspects. Individuals within the social enthusiasts and sceptic ascetics clusters are those living in the largest households who turn to SE more frequently, probably due to the intense social environment they are used to living in that makes them more extroverted.

Motivations and barriers are also at the core of Hallem et al.'s (2020) qualitative investigation, with in-depth interviews with 19 potential and actual SE consumers. The interviews reveal the main drivers (economic, social, sustainable consumption and societal responsibility) and impediments (trust issues, technological issues, acceptability issues, convenience issues, perceived needs issues) perceived by respondents, through which four different consumers' profiles have been delineated. Committed collaborative consumers turn to SE because they see it as a way of distancing themselves from the mainstream consumption model. Their sensibility to environmental issues leads them to choose SE for its potential to reduce their need for consumption. Moreover, they recognize in SE the opportunity to establish relations with other community members they share a lifestyle with. Pragmatist collaborative consumers are mostly driven by self-interest and utilitarianism; they thus see SE as a way to reduce expenditure while increasing the benefits and fun of consuming. As for their lower environmental concerns compared to the previous clusters, the possibility of adopting a more sustainable consumption mode is seen as a positive side effect rather than a key driver. Intermittent collaborative consumers turn to SE for its convenience and the possibility of performing quick and effective purchases. This is also the reason for their limited use of SE, when buying new products is more convenient. Thus, this cluster has no deep commitment to SE and is not interested in community belonging. Sceptical collaborative consumers exhibit a kind of resistance to SE because of the changes in habits it implies. They are not against it but are particularly attached to their traditional way of living and consuming; they do not see any necessity in engaging with SE while not excluding the possibility of future use.

Sands et al. (2020) have recourse to Social Exchange Theory (SET), which has often been used to study the approaches to SE (Grieco & Palagonia, 2024). According to this theory, social behaviour depends on the subjective evaluation each party carries out regarding the costs and benefits potentially derived from a resource exchange, with the ultimate goal of maximizing the benefits (Cropanzano & Mitchell, 2005). In their study, the authors analyse the heterogeneity of SE consumers, profiling segments based on different perceptions of possible benefits and risks of SE, and on the extent to which social norms play a role in influencing future sharing intentions. Four clusters emerge from their analysis. The mobility-focused sharers, as suggested by the label, are involved in the use of SE platforms within one specific industry, that is, often mobility. Compared with other clusters, individuals are older, have a lower level of innovativeness, do not believe that sharing reflects their social identity, and show a moderate intention towards the future use of SE. The diverse platform sharers adopt SE in more industries, report a high level of innovativeness and future sharing intention, and perceive SE reflecting considerably their social identity. A very similar profile is that of the power platform sharers who have the highest level of innovativeness and adopt SE across different industries, including retail and finance. The last segment is made up of non-sharers; they consistently have the lowest level of future intention, innovativeness, and a perception of sharing as reflecting their social identity. It is worth noting how all the four clusters recognize the potential unsuitability of the required product or service as the main risk associated with SE. The research also points to the different impacts that behavioural nudging might exert on the clusters, with no effect on three clusters out of four (non-sharers, mobility-focused sharers, and power platform sharers), and only one segment of individuals where intended behaviour could be altered.

Further studies classified SE consumers within specific industries. It is interesting to note that while the general classifications presented above adopt the demand perspective, starting from individuals' features and motivations, industry-specific investigations also adopt the offer perspective, showing how service configurations appeal to different segments of users. In the mobility sector, for example, Münzel et al. (2019) delved into the differences between consumers of B2C and P2P car sharing systems. In the accommodation industry, Lutz and Newlands (2018) highlight how the renters of a shared room differ from those renting entire homes in terms of both demographics and behavioural criteria. Guttentag et al. (2017) identified five motivating factors of Airbnb users, namely, interaction, home benefits, novelty, sharing economy ethos, and local authenticity, from which five different segments are derived (money savers, home seekers, collaborative consumers, pragmatic novelty seekers, and interactive novelty seekers). Focusing on toy library members, Ozanne and Ballantine (2010) clustered consumers according to the importance anti-consumption motives have in shaping their behaviour. What emerges is that there are nine constructs of consumer preference (friendship, sense of belonging, sense of duty, anti-consumption, parental mediating, frugality, materialism, toy library efficacy, and sharing), based on which four clusters of users can be identified: socialites, market avoiders, quiet anti-consumers, and passive members. Semaan et al. (2024) clustered the users of luxury sharing services, which have transformed the way consumers access and experience luxury, allowing for its democratization and higher accessibility in several different sectors. This led to the emergence of new consumer segments, driven by specific attitudes and motivations that, according to the authors, can be distinguished into four clusters: luxury indulgers, excellence-driven tradition alists, value seekers, and community engagers (Table 3.3).

Table 3.3 An overview of contributions about sharing economy consumers' clusters

Author/s (Year)	Theoretical framework	Sample size and location	Industries	Method for clustering
Hellwig et al. (2015)	Self-determination theory	10 SE users and 1121 both SE users and non-users (Germany and Switzerland)	Not specified	Semi-structured interviews + online self-administered survey
Lawson et al. (2016)	/	72 both SE users and non-users and 220 both SE users and non-users (US)	Not specified	Open-ended survey + online self-administered survey
Neunhoeffer and Teubner (2018)	/	745 both SE users and non-users (Germany)	Not specified	Online self-administered survey
Hallem et al. (2020)	/	19 both SE users and non-users	Not specified	In-depth semi-structured interviews
Sands et al. (2020)	Social exchange theory	411 both SE users and non-users (USA)	Mobility, retail, tourism, finance	Online self-administered survey
Münzel et al. (2019)	/	1835 both SE users and non-users (Netherlands)	Mobility (car sharing)	Online self-administered survey
Lutz and Newlands (2018)	/	699 Airbnb users and 50 Airbnb listings (USA)	Accommodation	Online self-administered survey + qualitative content analysis of Airbnb listings
Guttentag et al. (2017)	Innovation adoption	844 Airbnb users	Accommodation	Online self-administered survey
Ozanne and Ballantine (2010)	/	397 toy libraries members (New Zealand)	Toy libraries	Online self-administered survey
Semaan et al. (2024)	/	805 shared luxury service users (USA)	Luxury (accommodation)	Online self-administered survey

Source: Author's elaboration

3.4 Resistance to the Sharing Economy

3.4.1 A Look at Non-users

The above-mentioned contributions aim to cluster existing typologies of SE consumers, pointing to the elements that could define different user profiles. Yet, at the same time, they allow us to identify the key elements of non-users, who, despite being differently labelled (i.e. sharing opponents, fickle floaters, individualistic refuseniks, and non-sharers), show some common aspects (Table 3.4). To outline the non-user profile is of interest for academic and managerial research, even more so because this aspect has hitherto received scant attention. Extant literature mostly focuses on examining the factors that influence consumer participation in SE, leaving space for investigations into why consumers do not participate (Kansal & Bhalla, 2022). From the managerial point of view, although SE has undergone great expansion and is expected to continue to grow, its actual users are a small minority (Huynh & Gurtner, 2023).

What emerges is that non-users have a strong economic consciousness and are mostly driven by extrinsic motivation. They are not interested in signalling their social status through the use of SE platforms and, consequently, they do not associate their social identity with these alternative consumption modes. Risk seeking is not something that appeals to them, nor does the search for innovativeness. Conversely, they fear losing the object they own or receiving it back in an imperfect state, and their willingness to give is strongly tied to the expectation of having an equivalent value on return. From a demographic point of view, older and male individuals more often make up the non-user segments, in contrast with frequent users that usually present higher percentages of female and younger individuals. Household size also plays a role, with non-user segments belonging to the smallest.

It is important to note that, despite characterizing the profile of non-users, these aspects are not necessarily statistically significant in explaining SE (non) adoption. In fact, comparing users and non-users, Kansal and Bhalla (2022) found that economic reasons and materialism do not explain behavioural differences in SE adoption; the critical elements are institution- and reason-based trust. Huynh and Gurtner (2023) highlight the qualitative difference between reasons for SE adoption and reasons against. Their point is that understanding why consumers adopt SE does not imply understanding why they reject it, raising the need for ad hoc investigations.

3.4.2 The Main Barriers

Analysis of the barriers to SE goes along with the investigation into its drivers. Plenty of studies present statistical models of the impact motivations and constraints have on attitudes to SE, the intention to use it, and its actual adoption. Despite some elements recurring, the reasons preventing users from fully engaging with SE

Table 3.4 The features of non-users according to the identified clusters

Frequent usage	Average usage		Low or no usage
Sharing idealists [F, Pt]	**Sharing normatives**	**Sharing pragmatists** [M, Shs, Ft]	**Sharing opponents** [M, Ft]
Intrinsic motivations	Introjected motivations	Introjected motivations	Extrinsic motivations
+ Generosity General reciprocity – Perfectionism	+ Social media usage Perceived resource scarcity	– Generosity General reciprocity	+ Perfectionism Tit-for-tat reciprocity – Social media usage
Change seekers [Lhs, Hi, He]	**Conscious materialists** [Li, Le]	**Premium keepers**	**Fickle floaters** [O]
+ Seek variety – Possessiveness Materialism	+ Economic consciousness Environmental consciousness Possessiveness Materialism – Brand loyalty	+ Status consumption Brand loyalty – Economic consciousness	+ Economic consciousness – Status search Brand loyalty
Committed	**Pragmatist**	**Intermittent**	**Sceptical**
+ Anti-consumption Environmental consciousness Community belonging	+ Utilitarianism – Environmental consciousness	+ Convenience – Community belonging	+ Traditional consumption
Social enthusiasts [F, Lhs, Y]	**Sceptic ascetics** [Lhs]	**Conflicted materialists**	**Individualistic refuseniks** [M, Shs]
+ Risk seeking	– Ownership-related aspects Car owners + Risk seeking	+ Ownership-related aspects – Risk seeking	– Risk seeking
Diverse platform sharers	**Power platform sharers** [Y]	**Mobility-focused sharers** [O]	**Non-sharers**
+ Social identity Innovativeness Sustainable lifestyle Behavioural nudging	+ Social identity Innovativeness – Behavioural nudging	+ Ability to test new product – Innovativeness Behavioural nudging	– Innovativeness Social identity Behavioural nudging

Legend for demographic characteristics: *M* male; *F* = female; *Y* = young; *O* = old; *Hi* = high income; *Li* = low income; *He* = high education; *Le* = low education; *Lhs* = large household size; *Shs* = small household size; *Fu* = full-time worker, entrepreneur, or manager; *Pt* = part-time worker or homemaker

Source: Author's elaboration

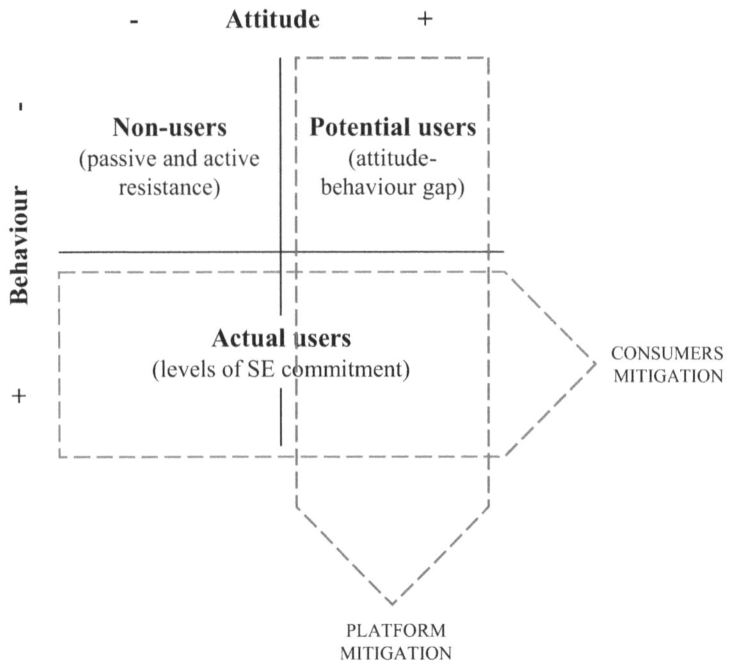

Fig. 3.3 The barriers to sharing economy: four conditions. Source: Author's elaboration

impact differently on the adoption process depending on the attitude and the behaviour of individuals (Fig. 3.3). Consumers with both a negative attitude and behaviour towards SE exhibit a strong SE resistance that, turning to innovation adoption studies, can be passive or active (Heidenreich & Handrich, 2015). Extant studies of SE barriers often focus on the latter, which depends on an unfavourable evaluation of the innovation. Conversely, passive resistance does not depend on the specific innovation and its evaluation; rather, it involves the willingness of individuals to accept the changes it brings and/or to break with their well-established norms and habits. Following this distinction, Huynh and Gurtner (2023) shed light on the passive resistance to SE that might be due to a general aversion to change or to satisfaction with the status quo. What emerges is that non-users perceive SE as impacting on their established habits and consumption patterns, and the associated behavioural change might prevent them from forming a positive attitude and, consequently, a positive behaviour to SE, even if this does not depend on a negative evaluation of SE itself. In their study, passive resistance can be conceptualized into adopter-specific and situation-specific factors. The first refers to individual inclination to resist changes and might depend on several reasons such as the reluctance to lose control, the lack of resilience, or a low tolerance for adjustment. The seconds concerns both preference for the prevailing situation and the psychological attachment consumers have to their objects. Active resistance includes (i) usage barriers, linked to strong habits that SE would affect; (ii) trust barriers, mostly due to the lack of professional providers; (iii) economic risk, associated with the perception of not spending money wisely; (iv) functional risk, mostly referred to product availability issues; (v) tradition barriers, depending on the perceived level of cultural change required by SE usage; and (vi) image barriers, relating to the ways in which SE is perceived and thus the impact it has on consumers' self-image. As for this last element, the shared use of objects with strangers can have different meanings depending on the characteristics of the users. For low-income individuals it could be a further mark of their exclusion from traditional purchasing channels, while for affluent segments it witnesses the choice of responsible consumption behaviour (Philip et al., 2015).

In most cases, the barriers to SE are investigated with potential users to assess which conditions prevent them from using these platforms, despite displaying a positive attitude. In fact, a positive attitude does not automatically translate into meaningful behaviour; this underscores the existence of the so-called attitude–behaviour gap. Hazée et al. (2019) stress the role of contamination concerns arising from worries about other people transferring residue or germs through touching objects. This aspect takes on a pivotal role in SE involving shared access to objects between unknown users, and the authors have demonstrated how it impedes adoption, especially for objects used in proximity to consumers' bodies. A similar finding comes from Edbring et al.'s (2016) investigation of the furniture and home products industry, where the perceived lack of hygiene limits second-hand purchase and the renting and sharing of goods, mostly for those product groups that are intimate to the users and/or hard to clean. The authors also shed light on the desire to own as another key deterrent to adoption, sometimes associated with the lack of trust in the other members of the sharing community (Edbring et al., 2016). The same emerges in the clothing industry, where the sense of ownership and the possession-self bond affect consumers' perceived risks and thus their willingness to engage in non-ownership consumption practices (Park & Armstrong, 2019). This happens to a lesser extent when redistribution of goods occurs through second-hand purchases, as consumers do not have to remove their ownership from the products. Park and Armstrong (2019) explain this through the endowment effect, according to which consumers' valuation of an object increases through its ownership. It is interesting to note that the relationship that individuals might establish with the objects they own does not always depend on their financial value. Hellwig et al. (2015) stress how individuals are more likely to share objects they have not incorporated into their self-identity, regardless of the objective value they have. This means that the extent to which ownership can hamper the sharing process is subjective, depending on which objects are more important as parts of an individual's extended self (Belk, 1988).

Barriers may also be experienced by actual users who have a positive attitude and behaviour towards SE. In this sense, the barriers do not prevent consumers from turning to SE, yet they determine some constraints to its use. Bhalla (2023) focuses on the factors that affect attitude formation for those users who exhibit a positive behaviour, revealing the critical role of materialism and perceived risk. This implies that individuals who are less inclined towards materialistic possession and do not perceive high performance and financial risks will have a positive attitude towards SE and, in turn, an increasing behavioural intention. Hazée et al. (2017, 2020) distinguish two categories of barriers that might also be experienced by actual users: functional and psychological. The former reflects the complexity and reliability of usage, where SE is difficult to understand or use, and consumers perceive uncertainty about its performance. The latter includes (i) compatibility with users' previous experience, social norms, usage patterns, or lifestyle; (ii) positive or negative image associations; (iii) the fear of contamination issues; and (iv) the concern about being held responsible for others' misuse of the shared asset.

Certainly, frequent usage has an impact on the barriers, with prior experience with SE lowering their perception (Ha et al., 2019; Shao et al., 2020). This is because consumers already using the service have gained the requisite knowledge and skills. Kurisu et al. (2021) carried out a comparative analysis between Japan and Switzerland, testing the impact of specific motivations and barriers on a sample of individuals grouped by how frequently they use SE services. What they discovered is that consumers with a greater familiarity perceive barriers that are more realistic, such as price and safety issues, while consumers unfamiliar with SE are not able to imagine the actual difficulties and tend to report more abstract negative impressions such as lack of interest or the perception that products are not available (Kurisu et al., 2021). Yet, consumers that are familiar with SE could still be reluctant to adopt it in specific circumstances, where its adoption might generate a sort of shame. Magnani and Re (2020) unveil this paradox in the car sharing realm, interviewing young adults about their lived experience with this service in two major cities (Milan and Amsterdam). Their empirical investigation revealed how although respondents declared to be enthusiast users of car sharing because of its functionality, efficiency, and innovativeness, they still feel shameful to adopt it when they feel exposed to the community's judgement (Magnani & Re, 2020).

Interestingly, scholars also shed light on the measures that might be adopted to mitigate the perception of these barriers, both at consumer and platform level. Hazée et al. (2020) investigate the practices that consumers adopt to attenuate the negative effects of these constraints when using SE. They propose that these practices decrease the impact of barriers on SE rejection when consumers perceive them as being easily doable. Based on their study, consumers could (i) distance themselves and ignore the critical aspect; (ii) adapt and self-organize to resolve the issue with their own resources; (iii) elaborate and reason to attenuate the constraints; (iv) exert social control on other users' behaviour and complain to the platform; or (v) engage in cooperating with the platform.

Mitigation measures may also come from the platform. Focusing on car sharing, Park and Armstrong (2019), for example, found how the configuration of some dimensions of the platform's business model (i.e. service level, pricing model, and type of product) can increase service usage and bridge the attitude-behaviour gap. Platform role is also important to attenuate contamination concerns which decrease with a high level of providers' brand equity, as consumers will have stereotype-related perceptions of the competence of other users. Moreover, appropriate placing of advertising messages could help, if the content does not highlight the physical contact between objects and users (Hazée et al., 2019). In fact, while information provision could be a measure to mitigate perceived barriers, stressing aspects such as cleaning certifications can paradoxically have a negative impact on consumers as it reminds them of possible contamination by others (Kurisu et al., 2021).

Focus: The Perspective of Providers

Among the actors in the SE relevant ecosystem, providers have been investigated to a lesser extent compared to the final consumers. Yet it is important to distinguish between consumers and providers when dealing with drivers and barriers since they have different roles and experience risks and impacts differently. Despite being beyond the scope of this book, it is interesting to note that providers' motives for renting out their possessions to peers are primarily economic, with the possibility of maximizing their revenue (Philip et al., 2015). Focusing on the barriers, functional risks act as strong restraints on providers engaging with SE; by "functional risks" is meant that they are worried about how consumers might use their items and that they risk them not being returned on time and in their original state (Huynh & Gurtner, 2023).

References

Aguilera-García, Á., Gomez, J., Sobrino, N., & Vinagre Díaz, J. J. (2021). Moped scooter sharing: citizens' perceptions, users' behavior, and implications for urban mobility. *Sustainability, 13*(12), 6886.

Albinsson, P. A., Perera, B. Y., Nafees, L., & Burman, B. (2019). Collaborative consumption usage in the US and India: An exploratory study. *Journal of Marketing Theory and Practice, 27*(4), 390–412.

Alzamora-Ruiz, J., Guerrero-Medina, C., Martínez-Fiestas, M., & Serida-Nishimura, J. (2020). Why people participate in collaborative consumption: An exploratory study of motivating factors in a Latin American economy. *Sustainability, 12*(5), 1936.

Bardhi, F., & Eckhardt, G. M. (2012). Access-based consumption: The case of car sharing. *Journal of Consumer Research, 39*(4), 881–898.

Belk, R. W. (1988). Possessions and the extended self. *Journal of Consumer Research, 15*(2), 139–168.

Bellotti, V., Ambard, A., Turner, D., Gossmann, C., Demkova, K., & Carroll, J. M. (2015, April). A muddle of models of motivation for using peer-to-peer economy systems. In *Proceedings of the 33rd annual ACM conference on human factors in computing systems* (pp. 1085–1094).

Berg, L., Slettemeås, D., Kjørstad, I., & Rosenberg, T. G. (2020). Trust and the don'twant-to-complain bias in peer-to-peer platform markets. *International Journal of Consumer Studies*, 44(3), 220–231.

Bhalla, S. (2023). Motivations and constraints of collaborative consumption, testing the mediating role of attitude and nature of trust. *Vision, 27*(2), 189–201.

Böcker, L., & Meelen, T. (2017). Sharing for people, planet or profit? Analysing motivations for intended sharing economy participation. *Environmental Innovation and Societal Transitions, 23*, 28–39.

Borusiak, B., Szymkowiak, A., Lopez-Lluch, D. B., & Sanchez-Bravo, P. (2021). The role of environmental concern in explaining attitude towards second-hand shopping. *Entrepreneurial Business and Economics Review, 9*(2), 71–83.
Botsman, R. (2017). *Who can you trust?: How technology brought us together–and why it could drive us apart*. Penguin UK.
Burinskienė, A., Leonavičienė, E., Grybaitė, V., Lingaitienė, O., & Merkevičius, J. (2021). Core elements affecting sharing: Evidence from the United States. *Sustainability, 13*(7), 3943.
Castelfranchi, C., & Falcone, R. (2010). *Trust theory: A socio-cognitive and computational model* (Vol. 18). Wiley.
Chatterjee, D., Dandona, B., Mitra, A., & Giri, M. (2019). Airbnb in India: Comparison with hotels, and factors affecting purchase intentions. *International Journal of Culture, Tourism and Hospitality Research, 13*(4), 430–442.
Chen, S.-C., & Quester, P. G. (2015). The relative contribution of love and trust towards customer loyalty. *Australasian Marketing Journal, 23*(1), 13–18.
Choudhury, V., & Karahanna, E. (2008). The relative advantage of electronic channels: A multidimensional view. *MIS Quarterly*, 179–200.
Cropanzano, R., & Mitchell, M. S. (2005). Social exchange theory: An interdisciplinary review. *Journal of Management, 31*(6), 874–900.
Dall Pizzol, H., Ordovás de Almeida, S., & do Couto Soares, M. (2017). Collaborative consumption: a proposed scale for measuring the construct applied to a carsharing setting. *Sustainability, 9*(5), 703.
Deci, E. L., & Ryan, R. M. (1985). *Intrinsic motivation and self-determination in human behaviour*. Plenum.
Deci, E. L., & Ryan, R. M. (2000). The "what" and "why" of goal pursuits: Human needs and the self-determination of behavior. *Psychological Inquiry, 11*(4), 227–268.
Deci, E. L., Koestner, R., & Ryan, R. M. (1999). A meta-analytic review of experiments examining the effects of extrinsic rewards on intrinsic motivation. *Psychological Bulletin, 125*(6), 627.
Edbring, E. G., Lehner, M., & Mont, O. (2016). Exploring consumer attitudes to alternative models of consumption: Motivations and barriers. *Journal of Cleaner Production, 123*, 5–15.
Ek Styvén, M., & Mariani, M. M. (2020). Understanding the intention to buy secondhand clothing on sharing economy platforms: The influence of sustainability, distance from the consumption system, and economic motivations. *Psychology and Marketing, 37*(5), 724–739.
Fan, A., Shin, H. W., Shi, J., & Wu, L. (2023). Young people share but do so differently: An empirical comparison of peer-to-peer accommodation consumption between millennials and generation Z. *Cornell Hospitality Quarterly, 64*(3), 322–337.
Grieco, C., & Palagonia, C. (2024). Delving into the behaviour of sharing economy consumers: A literature review. *Journal of Consumer Marketing, 41*(2), 162–179.
Grüner, A., Postel, L., & Schumann, J. H. (2024). Sharing is caring? The effect of negative peer-to-peer experiences on loyalty intentions in the sharing economy. *Journal of Business Research, 181*, 114757.
Grybaitė, V., & Stankevičienė, J. (2016). Motives for participation in the sharing economy–evidence from Lithuania. *Engineering Management in Production and Services, 8*(4), 7–17.
Guglielmetti Mugion, R., Toni, M., Di Pietro, L., Pasca, M. G., & Renzi, M. F. (2019). Understanding the antecedents of car sharing usage: An empirical study in Italy. *International Journal of Quality and Service Sciences, 11*(4), 523–541.
Guiot, D., & Roux, D. (2010). A second-hand shoppers' motivation scale: Antecedents, consequences, and implications for retailers. *Journal of Retailing, 86*(4), 355–371.
Guttentag, D., Smith, S., Potwarka, L., & Havitz, M. (2017). Why tourists choose Airbnb: A motivation-based segmentation study. *Journal of Travel Research*, 1–18.
Ha, J., Jung, H., Lee, H., & Choi, M. (2019). A study on the activation of car-sharing through the analysis of citizen's perception based on experience: Focusing on Busan, South Korea. *Urban Transport, 182*, 165–175.

Habibi, M. R., Davidson, A., & Laroche, M. (2017). What managers should know about the sharing economy. *Business Horizons, 60*(1), 113–121.

Hallem, Y., Ben Arfi, W., & Teulon, F. (2020). Exploring consumer attitudes to online collaborative consumption: A typology of collaborative consumer profiles. *Canadian Journal of Administrative Sciences/Revue Canadienne des Sciences de l'Administration, 37*(1), 82–94.

Hamari, J., Sjöklint, M., & Ukkonen, A. (2016). The sharing economy: Why people participate in collaborative consumption. *Journal of the Association for Information Science and Technology, 67*(9), 2047–2059.

Hartl, B., Sabitzer, T., Hofmann, E., & Penz, E. (2018). "Sustainability is a nice bonus" the role of sustainability in carsharing from a consumer perspective. *Journal of Cleaner Production, 202*, 88–100.

Hawlitschek, F., Teubner, T., & Gimpel, H. (2016, January). Understanding the sharing economy--Drivers and impediments for participation in peer-to-peer rental. In *2016 49th Hawaii International Conference on System Sciences (HICSS)* (pp. 4782–4791). IEEE.

Hawlitschek, F., Teubner, T., & Gimpel, H. (2018). Consumer motives for peer-to-peer sharing. *Journal of Cleaner Production*, 204, 144–157.

Hazée, S., Delcourt, C., & Van Vaerenbergh, Y. (2017). Burdens of access: Understanding customer barriers and barrier-attenuating practices in access-based services. *Journal of Service Research, 20*(4), 441–456.

Hazée, S., Van Vaerenbergh, Y., Delcourt, C., & Warlop, L. (2019). Sharing goods? Yuck, no! An investigation of consumers' contamination concerns about access-based services. *Journal of Service Research, 22*(3), 256–271.

Hazée, S., Zwienenberg, T. J., Van Vaerenbergh, Y., Faseur, T., Vandenberghe, A., & Keutgens, O. (2020). Why customers and peer service providers do not participate in collaborative consumption. *Journal of Service Management, 31*(3), 397–419.

Heidenreich, S., & Handrich, M. (2015). What about passive innovation resistance? Investigating adoption-related behavior from a resistance perspective. *Journal of Product Innovation Management, 32*(6), 878–903.

Hellwig, K., Morhart, F., Girardin, F., & Hauser, M. (2015). Exploring different types of sharing: A proposed segmentation of the market for "sharing" businesses. *Psychology and Marketing, 32*(9), 891–906.

Hinojo, P., Suárez, D., & García-Mariñoso, B. (2022). Drivers of consumer participation in online second-hand transactions. *Sustainability, 14*(7), 4318.

Hofmann, E., Hartl, B., & Penz, E. (2017). Power versus trust–what matters more in collaborative consumption? *Journal of Services Marketing, 31*(6), 589–603.

Huynh, T., & Gurtner, S. (2023). Resistance to the sharing economy: Why some consumers and providers do not participate in P2P sharing. *Journal of Cleaner Production, 422*, 138628.

Jain, S., & Mishra, S. (2020). Luxury fashion consumption in sharing economy: A study of Indian millennials. *Journal of Global Fashion Marketing, 11*(2), 171–189.

Jin, B., Sternquist, B., & Koh, A. (2003). Price as hedonic shopping. *Family and Consumer Sciences Research Journal, 31*(4), 378–402.

Kansal, P., & Bhalla, S. (2022). Collaborative consumption services: Testing the impact of reason-based trust and coercive power among users and non-users. *IUP Journal of Marketing Management, 21*(1).

Khalek, S. A., & Chakraborty, A. (2023). Access or collaboration? A typology of sharing economy. *Technological Forecasting and Social Change, 186*, 122121.

Khan, R., Awan, T. M., Fatima, T., & Javed, M. (2020). Driving forces of green consumption in sharing economy. *Management of Environmental Quality: An International Journal, 32*(1), 41–63.

Kim, N. L., & Jin, B. E. (2020). Why buy new when one can share? Exploring collaborative consumption motivations for consumer goods. *International Journal of Consumer Studies, 44*(2), 122–130.

Kim, D. J., Ferrin, D. L., & Rao, H. R. (2008). A trust-based consumer decision-making model in electronic commerce: The role of trust, perceived risk, and their antecedents. *Decision Support Systems, 44*(2), 544–564.
Kong, Y., Wang, Y., Hajli, S., & Featherman, M. (2020). In sharing economy we trust: Examining the effect of social and technical enablers on millennials' trust in sharing commerce. *Computers in Human Behavior, 108*, 105993.
Kurisu, K., Ikeuchi, R., Nakatani, J., & Moriguchi, Y. (2021). Consumers' motivations and barriers concerning various sharing services. *Journal of Cleaner Production, 308*, 127269.
Lamberton, C. P., & Rose, R. L. (2012). When is ours better than mine? A framework for understanding and altering participation in commercial sharing systems. *Journal of Marketing, 76*(4), 109–125.
Lang, B., Botha, E., Robertson, J., Kemper, J. A., Dolan, R., & Kietzmann, J. (2020). How to grow the sharing economy? Create prosumers! *Australasian Marketing Journal, 28*(3), 58–66.
Lawson, S. J., Gleim, M. R., Perren, R., & Hwang, J. (2016). Freedom from ownership: An exploration of access-based consumption. *Journal of Business Research, 69*(8), 2615–2623.
Lee, S., Lee, W., Vogt, C. A., & Zhang, Y. (2021). A comparative analysis of factors influencing millennial travellers' intentions to use ride-hailing. *Information Technology and Tourism, 23*(2), 133–157.
Lindenberg, S. (2001). Intrinsic motivation in a new light. *Kyklos, 54*(2–3), 317–342.
Lutz, C., & Newlands, G. (2018). Consumer segmentation within the sharing economy: The case of Airbnb. *Journal of Business Research, 88*, 187–196.
Magnani, G., & Re, B. (2020). Lived experiences about car sharing in young adults: Emerging paradoxes. *Italian Journal of Marketing, 2020*(2), 207–229.
Marth, S., Hartl, B., & Penz, E. (2022). Sharing on platforms: Reducing perceived risk for peer-to-peer platform consumers through trust-building and regulation. *Journal of Consumer Behaviour, 21*(6), 1255–1267.
Mayer, R. C., Davis, J. H., & Schoorman, F. D. (1995). An integrative model of organizational trust. *The Academy of Management Review, 20*(3), 709–734.
Merino-Saum, A., Jemio, P. R., Hansmann, R., & Binder, C. R. (2023). Drivers and barriers to participation in the sharing economy: Does the environment really matter? A systematic review of 175 scientific articles. *Resources, Conservation and Recycling, 198*, 107121.
Möhlmann, M. (2015). Collaborative consumption: Determinants of satisfaction and the likelihood of using a sharing economy option again. *Journal of Consumer Behaviour, 14*(3), 193–207.
Möhlmann, M., & Geissinger, A. (2018). Trust in the sharing economy: Platform-mediated peer trust. *Cambridge Handbook of the Law of the Sharing Economy, 70*(1), 26–44.
Münzel, K., Piscicelli, L., Boon, W., & Frenken, K. (2019). Different business models–different users? Uncovering the motives and characteristics of business-to-consumer and peer-to-peer carsharing adopters in The Netherlands. *Transportation Research Part D: Transport and Environment, 73*, 276–306.
Neunhoeffer, F., & Teubner, T. (2018). Between enthusiasm and refusal: A cluster analysis on consumer types and attitudes towards peer-to-peer sharing. *Journal of Consumer Behaviour, 17*(2), 221–236.
Ozanne, L. K., & Ballantine, P. W. (2010). Sharing as a form of anti-consumption? An examination of toy library users. *Journal of Consumer Behaviour, 9*(6), 485–498.
Park, H., & Armstrong, C. M. J. (2019). Will "no-ownership" work for apparel?: Implications for apparel retailers. *Journal of Retailing and Consumer Services, 47*, 66–73.
Philip, H. E., Ozanne, L. K., & Ballantine, P. W. (2015). Examining temporary disposition and acquisition in peer-to-peer renting. *Journal of Marketing Management, 31*(11–12), 1310–1332.
Piscicelli, L., Cooper, T., & Fisher, T. (2015). The role of values in collaborative consumption: Insights from a product-service system for lending and borrowing in the UK. *Journal of Cleaner Production, 97*, 21–29.

Ponte, E. B., Carvajal-Trujillo, E., & Escobar-Rodríguez, T. (2015). Influence of trust and perceived value on the intention to purchase travel online: Integrating the effects of assurance on trust antecedents. *Tourism Management, 47*, 286–302.

Rebiazina, V., Shalaeva, A., & Smirnova, M. (2018). Do Russian consumers understand and accept the sharing economy as a new digital business model?. In *Digital transformation and global society: Third international conference, DTGS 2018*, St. Petersburg, May 30–June 2, 2018, Revised Selected Papers, Part I 3 (pp. 386–401). Springer International Publishing.

Resnick, P., Kuwabara, K., Zeckhauser, R., & Friedman, E. (2000). Reputation systems. *Communications of the ACM, 43*(12), 45–48.

Sands, S., Ferraro, C., Campbell, C., Kietzmann, J., & Andonopoulos, V. V. (2020). Who shares? Profiling consumers in the sharing economy. *Australasian Marketing Journal, 28*(3), 22–33.

Schor, J. (2016). Debating the sharing economy. *Journal of self-governance and management economics*, 4(3), 7–22.

Semaan, R. W., Boukis, A., Christodoulides, G., & Papastathopoulos, A. (2024). A typology of consumers sharing luxury services. *Journal of Strategic Marketing, 32*(4), 536–550.

Serrano, L., Ariza-Montes, A., Nader, M., Sianes, A., & Law, R. (2021). Exploring preferences and sustainable attitudes of Airbnb green users in the review comments and ratings: A text mining approach. In *Sustainable consumer behaviour and the environment* (pp. 114–132). Routledge.

Shao, Z., Li, X., Guo, Y., & Zhang, L. (2020). Influence of service quality in sharing economy: Understanding customers' continuance intention of bicycle sharing. *Electronic Commerce Research and Applications, 40*, 100944.

Simpson, J. A. (2007). Psychological foundations of trust. *Current Directions in Psychological Science, 16*(5), 264–268.

Ter Huurne, M., Ronteltap, A., Corten, R., & Buskens, V. (2017). Antecedents of trust in the sharing economy: A systematic review. *Journal of Consumer Behaviour, 16*(6), 485–498.

Tsai, Y. T., Yu, C. H., & Boonprakob, R. (2021). Assessing carpooling drivers and barriers: Evidence from Bangkok, Thailand. *Transportation Research Part F: Traffic Psychology and Behaviour, 82*, 84–95.

Tunçel, N., & Özkan Tektaş, Ö. (2020). Intrinsic motivators of collaborative consumption: A study of accommodation rental services. *International Journal of Consumer Studies, 44*(6), 616–628.

Violin, F. L. (2021). Motivators for the use of vehicle sharing services on demand in Brazil. *Journal of Globalization, Competitiveness and Governability, 15*(3).

Wood, J. A., Boles, J. S., & Babin, B. J. (2008). The formation of buyer's trust of the seller in an initial sales encounter. *Journal of Marketing Theory and Practice, 16*(1), 27–39.

Wu, J., Zeng, M., & Xie, K. L. (2017). Chinese travelers' behavioral intentions toward room-sharing platforms: The influence of motivations, perceived trust, and past experience. *International Journal of Contemporary Hospitality Management, 29*(10), 2688–2707.

Xu, X., Zeng, S., & He, Y. (2021). The impact of information disclosure on consumer purchase behavior on sharing economy platform Airbnb. *International Journal of Production Economics, 231*, 107846.

Zhang, T., Bufquin, D., & Lu, C. (2019). A qualitative investigation of micro-entrepreneurship in the sharing economy. *International Journal of Hospitality Management, 79*, 148–157.

Chapter 4
"Happily Ever After"

The Post-purchase Behaviour of Sharing Economy Consumers

4.1 Evaluating Sharing Economy Quality

The perception consumers have about SE services in the post-adoption phase has a pivotal role in defining the success of platforms. This is as important as understanding consumer motivations. A positive evaluation of their experience will determine consumers' satisfaction and, in turn, result in repurchases and loyal behaviours. When focusing on post-adoption behaviour, the starting point is evaluating the quality of the service received, a milestone concept in marketing sciences (Bolton & Drew, 1991). Service quality refers to the expost appraisal of the experience based on the ex ante expectations consumers developed. In this context, three different conditions are possible, depending on whether the experience falls below, matches, or exceeds expectations. In the first case, the difference between the expected and experienced service will be negative, leading consumers to be dissatisfied. In the second and third cases, the difference will be positive, and consumers will be satisfied or delighted respectively (Parasuraman et al., 1985). A key issue for practitioners is to understand the dimensions by which consumers form their expectations and evaluate services, with a view to offering experiences that meet or exceed them. This is very important for developing retention strategies. Nevertheless, while existing research largely points to the importance of the service quality concept, studies

C. Grieco, *Consumer Behavior in the Sharing Economy*, Contributions to Management Science, https://doi.org/10.1007/978-3-031-76279-6_4

investigating its specific features in SE are few and far between (Shao et al., 2020). Studies into consumer behaviour in SE are often focused on initial intention, leaving other types of intention, i.e. continuance intention, poorly covered (Wang et al., 2021).

The theoretical foundation of service quality lies in the expectancy-disconfirmation theory, first developed by Oliver (1980), studying customer purchase intentions, and later extended by Kristensen et al. (1999) into expectation-confirmation theory (ECT). According to this theory, consumers develop a certain level of expectation about a product or a service before purchase. Once the purchase has been made, consumers form their perception of performance confirming or disconfirming their prior expectations. High levels of confirmation positively impact on how useful consumers perceive the service to have been, triggering satisfaction and repurchase intentions (Bhattacherjee, 2001a, b). This theoretical model has been widely investigated with online exchanges. For example, Kim et al. (2009) focused on post-purchase behaviour in e-commerce, showing how trust has a long-term impact on reinforcing confirmation and, in turn, e-loyalty. Lin et al. (2005) studied users' continued use of websites and integrated the perceived playfulness in the ECT framework as a further variable in fostering repurchase, along with perceived usefulness. Wang et al. (2021) demonstrated how ECT supports the explanation about consumers repurchase intentions, yet fails to explain why they follow different paths, such as switching to other products or services. In this sense, they integrate the ECT framework, going beyond the dissatisfaction that might come from unmet expectations, and focus on possible changes in consumer attitudes and behaviour towards services. The authors also move forward in the satisfaction case, integrating recommendation intention as a further positive outcome potentially derived from confirmation (Wang et al., 2021).

4.1.1 Assessment Models for Service Quality

From the platform perspective, a full understanding of service quality is essential, given its strategic importance. This has led scholars to develop a wide range of scales and metrics to measure the extent to which consumer expectations have been met by the service provided. A first distinction needs to be made between incident-based and attribute-based methods (Stauss & Weinlich, 1997). The former considers an event to be either satisfactory or unsatisfactory. The latter conceptualizes service quality as multidimensional and hence affected by several different elements (Priporas et al., 2017; Yoo & Yoon, 2022). Service quality thus needs to be evaluated through an attribute-level approach, as consumers may have a positive perception of one dimension and a negative perception of another (Ahrholdt et al., 2017). A further distinction can be made between qualitative and quantitative models, where the former visually represents a framework describing the relations between factors, and the second provides a collection of items to obtain a score that measures service quality (Amat-Lefort et al., 2020). The best known and most adopted measurement tool has been developed by Parasuraman et al. (1985, 1988) and comprises both. The qualitative framework is called the GAP service quality model and based on ECT explains how perceived quality depends on the ability of the service provider to avoid gaps forming between consumer expectations and experiences. The model identifies five areas of possible gaps:

- *Knowledge gap*: the service provider has poor or false understanding of what consumers expect from the service.
- *Design Gap*: the service provider does not design the service according to expectations.
- *Performance gap*: the service provider is not able to provide the service standards designed for the service.
- *Communication gap*: what the service provider says about the service does not match service delivery.
- *Consumer gap*: the consequence of any of the four gaps listed above is that consumers perceive a difference between the expected service and the actual experience.

A quantitative model called SERVQUAL (SERvice QUALity model) has consequently been developed to measure service quality based on five dimensions: tangibles, reliability, responsiveness, assurance, and empathy. The SERVQUAL model has been largely adopted and adapted to specific domains. An important update comes from its adaptation to online services resulting from the development of the Internet. To this end, Parasuraman et al. (2005) developed the E-S-QUAL, a model consisting of 22 items grouped into four factors: efficiency of website interaction; fulfilment of the service promised; system availability and management of technical issues; and privacy protection. Further studies added dimensions such as website design, information quality, interface interactivity, and service failure recovery (Cheng et al., 2018).

Despite differences in the elements proposed, these service quality models are usually based on the gap between expectations and evaluation; they see the final perception of quality as resulting from a comparison between the service experienced and the prior considerations of consumers. Following a different path, the SERVPERF model developed by Cronin and Taylor (1992, 1994) posits that service quality is evaluated by assessing actual performance, without considering previous consumer expectations and experiences (Wireko-Gyebi et al., 2024) (Table 4.1).

4.1.2 The Need for Sharing Economy Tailored Models

The models described above have been largely adopted in traditional business and adapted to meet the specificities of different sectors. However, referring to SE, several studies have underlined the need for tailored models capable of fully capturing those key aspects that make SE platforms so different from conventional businesses. There are two elements of the SE business model that make this necessary. Firstly,

Table 4.1 Models for service quality in traditional businesses

Author/s (Year)	Model name	Dimensions
Grönroos (1982)	Service Quality Model	Technical quality, functional quality, corporate image
Parasuraman et al. (1985)	GAP Model	Reliability, responsiveness, competence, access, courtesy, communication, credibility, security, understanding/knowing the customer, tangibles
Haywood-Farmer (1988)	Service Quality Attributes	Physical facilities, processes and procedures, people behaviour and conviviality, professional judgement
Parasuraman et al. (1988)	SERVQUAL	Tangibles, reliability, responsiveness, assurance, empathy
Cronin and Taylor (1992)	SERVPERF	Same as SERVQUAL but with performance only statements
Knutson et al. (1990)	LODGSERV	Same as SERVQUAL but for hotels
Stevens et al. (1995)	DINESERV	Same as SERVQUAL but for restaurants
Dabholkar et al. (1996)	RSQS	Physical aspects, reliability, personal interaction, problem-solving, policy
Philip and Hazlett (1997)	PCP Model	Pivotal, core, peripheral attributes
Brady and Cronin Jr (2001)	Service Quality Model	Personal interaction quality, physical service environment quality, outcome quality
Yoo and Donthu (2001)	SITEQUAL	Competitive value, product quality assurance, clarity of ordering, ease of use, aesthetic design, processing speed, security, corporate and brand equity, product uniqueness
Loiacono et al. (2002)	WebQual	Online completeness, response time, trust, tailored communications, ease of understanding, intuitive operations, visual appeal, innovativeness, emotional appeal, consistent image, information fit-to-task, relative advantage
Parasuraman et al. (2005)	E-S-Qual	Efficiency, fulfilment, system availability, privacy
Parasuraman et al. (2005)	E-RecS-Qual	Responsiveness, compensation, contact

Source: Author's elaboration basing on Amat-Lefort et al. (2020)

the triadic relationship between users, providers, and platforms. In this regard, a quality model for SE needs to consider the interaction of consumers with these actors, both of which determine the overall perception of quality. Secondly, these interactions take place both online and offline, and these channels are used separately or simultaneously along the different phases of the consumer journey. Studies covering this are few and mostly based on the principle that the specific dimensions of service quality cannot be standardized, as they vary significantly between different settings. For this reason, the models they suggest for measuring service quality in SE are often framed within a specific industry, mostly accommodation and mobility (Fig. 4.1).

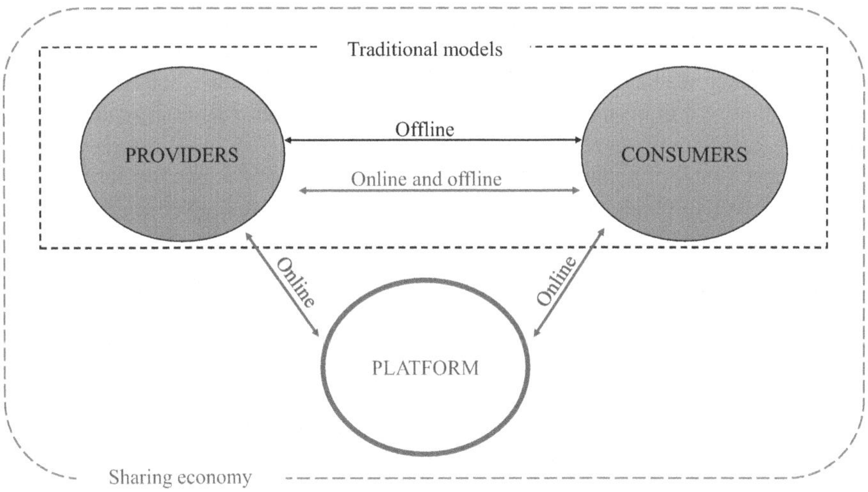

Fig. 4.1 The specificities of SE business model compared to traditional business. Source: Author's elaboration

As for the former, a first attempt comes from Priporas et al. (2017) who adopted a scale developed for traditional hotels to measure service quality as perceived by Airbnb consumers. Their study indicated how Airbnb consumers give particular attention to the convenience of the experience (i.e. access to specific tourist sights, ease of acquiring information, efficiency of problem resolution), assurance (i.e. the feeling of being safe and secure), and the courtesy of renters. On the other hand, tangibles and the adequacy of service supply are not as important as in the hotel industry (Priporas et al., 2017). Amat-Lefort et al. (2023) go a step further in proposing specific dimensions for SE service quality in the accommodation sector. In particular, they identify six dimensions: (i) comfort, which refers to the most basic yet vital features of accommodation services, such as cleanliness; (ii) social interaction, indicating the friendly contact with the host and/or other guests in shared accommodation; (iii) legal protection, relating to the protection given to guests for any potential issue; (iv) platform responsiveness, i.e. the capacity of the platform to react promptly to guests' requests; (v) site organization, making the platform easy to use in terms of navigation and accessibility; and (vi) trust, that is, degree of platform trustworthiness. According to their view, SE needs an ad hoc measurement scale for service quality, as none of the existing models consider the continuous interaction consumers have with the digital platform, and the role of trust as a prerequisite in SE transactions. These are the key original elements of their study compared to previous contributions. In studying Couchsurfing, Bahadur and Ali (2023) point to the importance of fostering emotional connections in SE business models and demonstrate how a positive evaluation in this dimension impacts on both consumer satisfaction and subjective well-being. Yoo and Yoon (2022) analysed six features of accommodation to compare the importance given by Airbnb guests versus traditional hotel travellers. Their findings reveal that Airbnb users put more weight on the presence of a kitchen or kitchenette, staff availability, ratings, and aesthetic design when evaluating the service quality of their stays. Conversely, they are less interested in room cleanliness and size, both of which are of greater importance for travellers using traditional hotels (Yoo & Yoon, 2022). The authors interpreted these results by stating that Airbnb accommodation is usually larger, as it mostly consists of residential properties. For this reason, users take for granted that they will have enough space, while in traditional hotels the rented room is the only private lodging space. The presence of kitchen or kitchenette adds considerable value because they make the Airbnb solution more economical compared to traditional hotels. Attention to other services such as staff availability is closely linked to the trust issue, as consumers may need to be reassured about their safety when renting a space that is privately owned or shared with other guests. All six identified elements influence satisfaction and purchase intention in both sharing and traditional services (Yoo & Yoon, 2022).

In the mobility industry, Cheng et al. (2018) studied car-hailing services in China, providing a measurement scale that distinguishes between factors in online and offline service quality. As for the former, structural assurance and platform responsiveness were found to be important, while for the latter a key role is played by information congruity, competence, and empathy. Furthermore, the authors point to the importance of a positive attitude as a key element in influencing the perception of quality. A highly positive attitude towards SE implies a high level of perceived quality, both for offline and online interactions and, in turn, high levels of loyalty (Cheng et al., 2018). Shao et al. (2020) focus on bike sharing and propose a re-conceptualization of the SERVQUAL model appropriate to the specificities of dockless services. In their view, traditional elements of the framework need to be adapted to the features of this business model. For example, reliability needs to include the provision of real-time information about vehicle availability based on previous user feedback; empathy also takes on a strategic meaning, as this service involves recording users' preferences and offering them personalized notifications; assurance covers all issues relating to the transactions, whether monetary or data. Responsiveness and tangibles also need to be taken into consideration in both online and offline configurations of the service (Shao et al., 2020). Their research has shed light on the impact all five of these dimensions have on consumer confirmation while also revealing some gender-based differences. It appears that male users tend to focus more on location reliability and striking appearance, while female users pay more attention to transaction security. This provides evidence of the high level of subjectivity characterizing evaluation.

Although service quality is hard to standardize, the focus on a specific sector, if not a specific platform, can limit the development of this field of study. In this context, scholars have worked on developing systems of evaluation that can take into account SE specificities regardless of the industry. Referring to the distinction between qualitative and quantitative models, Amat-Lefort et al. (2020) and Marimon et al. (2019) have provided two notable studies covering the first and second types of contribution, respectively. Amat-Lefort et al. (2020) adapted the GAP service model in suggesting a conceptual framework which considers the interactions SE consumers have with both providers and platforms. The authors reshape Parasuraman's framework into a cyclical process where each of the identified gaps can be seen as a step leading to the next one, resulting in a disjunction between consumer perceptions and expectations. Moreover, they identify two different cycles, where the same gaps refer to the two types of interactions consumers have. Interestingly, the authors clarify how these two cycles include phases that may be separate, overlapping, or intersecting, depending on the phases of the consumer's journey. If the platform is incapable of capturing satisfactorily what consumers expect from the service, it will fail to meet their desires; it is likely that the associated providers will do the same (Amat-Lefort et al., 2020).

Marimon et al. (2019) focus on the need to develop a quantitative model, giving weight to the differences between online and offline interactions. In this context, they analyse existing contributions on offline and online service quality models and claim that merging the two inadequately reflects the complexity of SE. The model they propose is called CC-Qual (Collaborative Consumption Quality) and comprises 21 items grouped into five dimensions. The first three refer to online interactions: (i) site organization relates to the way information is displayed and accessible on the website; (ii) platform responsiveness and agility focuses on the quality of platform intervention in case of issues; and (iii) legal protection and trustworthiness reflects consumer concerns about the security of transactions. These three dimensions focus on the relationship between consumers and platforms, as their encounter is entirely online. A further dimension refers to the relationship with the provider; labelled (iv) interaction with providers, it examines their competence, empathy, and trustworthiness. A last dimension is (v) social interaction and covers the encounter with other users, as in the case of Airbnb or Blablacar where more individuals might share the asset together. As interactions with other individuals may take place either online or offline, these two last dimensions incorporate the evaluation of both these elements (Marimon et al., 2019).

4.2 Customer Satisfaction in the Sharing Economy

According to ECT, perception that a service has met or exceeded their prior expectations determines the extent to which consumers are (dis)satisfied with their experience. Satisfaction is thus the overall cognitive and affective response to the service; it is essential for platform growth because it is linked to repurchase intention. Satisfied consumers perceive that their needs have been fulfilled and that the benefits they gain from the purchase exceed the costs involved (Zhou & Zhang, 2019). From the platform perspective, its recognition indicates that the service offered satisfactorily meets consumers' desires (Phuong & Trang, 2018). The link between service quality and consumer satisfaction has been extensively assessed, producing different views on the impact of the former on the latter. Existing contributions mainly consider service quality to be an antecedent of satisfaction, both overall

(Wireko-Gyebi et al., 2024) and when focusing on specific dimensions of the SERVQUAL model. For example, Kumar et al. (2022) found that reliability is the only dimension that notably affects the satisfaction of Uber's consumers. Ziyad et al. (2020) identify the bundle of assurance, empathy, and responsiveness as significantly contributing to satisfaction in car-sharing platforms. On the other hand, other studies disconfirm this relationship in SE, stating that the level of customer satisfaction has a regulatory effect on perceived service quality (Zhou & Zhang, 2019). Möhlmann (2015) compared car2go and Airbnb to reveal the ways in which the key drivers of SE impact on consumer satisfaction and repurchase intention in B2C and C2C services. She discovered that service quality only significantly determines satisfaction in the context of car sharing; in the accommodation industry it is irrelevant.

In studying Airbnb, Liang et al. (2018) refer to the distinction between transaction-based and experience-based satisfaction. The first sums up the extent to which consumers are satisfied with a specific service encounter, while the second concerns the overall experience with the platform. This means that consumers might also adopt different criteria in evaluating an aspect of the transaction and the overall experience, as they lead to related but distinct types of satisfaction. Their findings show that, for Airbnb consumers, transaction-based satisfaction is the more critical predictor of repurchasing behaviour and directly impacts on experience-based satisfaction. This difference points to an interesting aspect of SE, namely, the dual relationship consumers form with platform and providers, and thus to the different sources and types of satisfaction they can derive. Every time consumers book a stay on Airbnb, they deal with different hosts and different accommodation while using the same platform (Liang et al., 2018). Evaluation of the service offered by the peer provider plays a greater role in determining overall satisfaction, compared to that of the platform. Jin and Chen (2020) confirm this consideration in car-hailing services, explaining that the interaction with the peer provider (driver) adds value to the overall experience and is more important than relating to the platform in enhancing both satisfaction and loyalty. A further element impacting on the interaction between consumers and providers which increases overall perception of satisfaction is the level of knowledge that the parties have of each other before the transaction. Stough and Carter (2023) demonstrated this through an experimental study on ski renting. Their findings reveal that providing consumers with information about the previous users of products negatively impacts on their satisfaction with the product, unless the information refers to positive traits.

Studies focusing on the elements that determine satisfaction with SE are mostly based on the same aspects under investigation as its key drivers (e.g. SDT's intrinsic and extrinsic motivations). Hedonic value emerges as a key determinant of consumer satisfaction; this refers to the importance of experience enjoyment and pleasure (Tussyadiah, 2016). Convenience and economic value are also pivotal aspects (Cho, 2020), particularly in contexts such as car sharing, where consumers are aware that these services, being cheaper than other traditional modes, offer them the possibility of saving money (Möhlmann, 2015; Kaur et al., 2016; Yapp & Yeap, 2023). Focusing on mobility, further issues relating to payment are closely linked to overall satisfaction, for example, the possibility that consumers can know in advance the amount of money they will have to pay for the service (Chen & Sheldon, 2015); the ability to complete the payment in several ways both before and after the trip (Azmi et al., 2019); and the ease and safety of payments (Hughes & MacKenzie, 2016). Surprisingly, the search for a sustainable lifestyle has a negative effect on satisfaction. The research carried out by Tussyadiah (2016) on Airbnb consumers reveals how environmental concerns have scarce if any significance in determining satisfaction, while also questioning the role of sustainability as a reason driving consumers towards SE, at least in the accommodation industry. Social benefits also have limited affect, except for those renting private rooms in cohabitated spaces who are more inclined to appreciate interactions with the host or other users. Conversely, those renting an entire property are very likely trying to avoid much interaction; here, the emphasis on social benefits results in a less satisfying stay (Tussyadiah, 2016). The same result is also confirmed by Möhlmann (2015), who further stresses the impact of familiarity, trust, and utility on Airbnb consumers' satisfaction. The platform application itself has the potential to impact on satisfaction. In particular, consumers appreciate the structure of the interface and the interactive features that allow communication with other users and provide real-time updates and feedback (Rasheed & Nandukrishna, 2023).

4.2.1 The Consequences of (Dis)satisfied Consumers

The level of satisfaction determines the adoption of post-usage behaviours that could in turn be of relevance to other users, whether potential or actual consumers. Satisfied consumers are more likely to keep on using the service, making satisfaction a precondition for repurchase, both at attitudinal and behavioural levels (Arteaga-Sánchez et al., 2020).

Continuance intention may also involve adopting a new role in the SE exchange process. An interesting consideration is noted by Ertz et al. (2022) who stress the role of satisfaction as one of the elements potentially driving the users-to-providers transition. Their research reveals how high levels of satisfaction could lead consumers to continue using a platform not only as end users but as service providers, with strategic benefits for the platform's success. Providers have 3.5 times more impact than consumers in determining the growth of the business model (Chu & Manchanda, 2016).

Positive effects of satisfaction also include the likelihood of disseminating the worth of the service by word of mouth (WOM) (Tanford, 2016; Slack et al., 2020); this is a further strategic element in fostering platform success. Consumers recommending a product or service set up informal communication within their networks that could significantly influence others' consumer behaviour. The more satisfaction consumers experience, the greater will be their intention to recommend (Rasheed & Nandukrishna, 2023).

A related yet different consideration could apply to rating and reviewing systems. As mentioned above in the context of trust, peer-to-peer transactions lack the guarantees provided by a company brand, and for this reason, ratings and reviews are essential to ensure both consumers and providers about the safety and effectiveness of the exchange. SE platforms largely rely on these trust-building tools to protect parties from fraud and risks. These mechanisms allow both satisfied and dissatisfied consumers to express their opinion about the service; however, studies reveal two key issues that may hamper their full effectiveness. On the one hand, although checking reviews and ratings before a transaction is widespread, only a small portion of consumers give their evaluation afterwards (European Commission, 2017). On the other hand, consumers are also reluctant to give negative ratings and reviews publicly and are thus more inclined to avoid complaining even when there are valid reasons for doing so (Zervas et al., 2021). This aspect is well described by Berg et al. (2020) as the "don't want to complain" bias, evidenced and conceptualized in their research on trust in SE review systems. The authors point out that dissatisfied consumers rarely leave negative reviews because of this feeling of not wanting to complain, originating in the desire to act in a socially desirable way rather than purely rationally. The consequence is that this bias ends up undermining the review system, leading consumers to lose trust in mechanisms seen as not giving adequate information about providers and platforms (OECD, 2017). This response is aggravated by suspicions that the platform itself filters or moderates other users' online ratings (Akhmedova et al., 2020).

Dissatisfied consumers will be less likely to continue using the same SE service; instead, they will eventually manifest switching intentions. This means that if the performance is not satisfactory, consumers will look for other SE or non-SE options to meet their requirements (Wang et al., 2021). The switch intention stems from low satisfaction coupled with the existence of alternatives and the costs and benefits of switching (Ghazali et al., 2016). Given these conditions, dissatisfied consumers will see the alternatives as more attractive and able to offer better services, making it difficult for the platform to retain them, also taking into account the increasing number of SE alternatives (Jiang & Lau, 2021).

It is worth noting that satisfaction can play a key role in mitigating the negative effects of service failure. Users satisfied with their relationship with a platform manifest high trust and low risk perception (Hess Jr et al., 2003). In the case of a negative experience, consumers with a high level of satisfaction will align this with their positive preconceptions, thereby limiting any decrease in platform trust they would have otherwise experienced (Grüner et al., 2024). This is particularly true for outcome-related experiences (i.e. what customers receive from the service) compared to process-related experiences (i.e. how customers receive the service) (Grüner et al., 2024).

Focus: The Rebound Effect of the Sharing Economy

A dynamic related to post-purchase that impacts on the wider discourse on SE and sustainability is the so-called rebound effect. This concept falls within the negative side of SE in referring to a behaviour or other response to a measure that offsets the effect of the measure itself. It has been conceptualized in the context of energy conservation and occurs when the unit price of a good or service decreases due to an improvement in efficiency and, as a consequence, people demand more of it (Dütschke et al., 2018). Studies into the rebound effect of SE, hitherto few in number, have recently come out showing that it occurs mostly when consumers are not driven by environmental concerns but by economic and hedonistic reasons (Merino-Saum et al., 2023). Already existing contributions have so far focused on specific effects and/or specific product/service categories. For example, Ackermann and Tunn (2024) focus on the careless use of washing machines; Meshulam et al. (2023) highlight how the money saved through food sharing platforms is then spent to purchase more food via conventional channels. Other studies focus on mobility sharing whose impact on traffic congestion, pollution increase, and lower public transportation use has been proven (e.g. Nozari et al., 2021).

4.3 The Challenge of Retaining Users

The ultimate goal for a company is to generate loyalty in its consumers. Oliver (2014) defines customer loyalty as the "deeply held guarantee to re-buy or re-patronise a favoured product or service consistently in the future, despite situational influences and marketing efforts that have the potential to cause switching behaviours" (Oliver, 2014, p. 432). This not only means that they exhibit repurchase behaviour because of their satisfaction with the service, but also that they have developed a favourable attitude towards the brand and an emotional attachment to its products or services. Scholars distinguish between two different dimensions of loyalty: behavioural loyalty and attitudinal loyalty. The first dimension refers to consumers' repeated purchases without necessarily implying a psychological affiliation. A loyal behaviour alone is not sufficient to wholly explain the notion of loyalty, as it could be fake (Hawkins & Vel, 2013). Being behaviourally centred, this type of loyalty may change along with changes in a company's marketing strategy (Al-dweeri et al., 2017). The second dimension, attitudinal loyalty, concerns emotional attachment to the product or brand that leads consumers to repeat the purchase out of a favourable and solid attitude towards it (Anatolevena Anisimova, 2007). Attitudinal loyalty can be assessed by asking consumers about their intention to recommend the product or service to others (Lee & Wong, 2021).

Fostering loyal behaviour is pivotal for a company's success as there are several associated benefits. Loyalty leads to more purchases that increase familiarity with the brand while reducing perception of risk. Furthermore, price sensitiveness is reduced, and the tendency to spread positive WOM grows. Loyal consumers also reduce the costs of acquisitions and increase a company's profitability (Kumar & Shah, 2004).

Customer loyalty largely depends on customer satisfaction which originates in a positive evaluation of service quality. These nexuses do not imply that one element would automatically lead to the others, yet each step is a necessary precondition of the following (Tanford, 2016; Yoo & Yoon, 2022). Thus, despite not all satisfied consumers becoming loyal, satisfaction is a crucial precondition of loyalty. The same applies to service quality. It does not automatically translate into satisfaction but is self-evidently a key determinant of it. As for the connection between these elements, the psychological bond that consumers establish with the product or service from the initial purchase is fundamental in providing the foundations of future attitudinal and behavioural loyalty (Bakti & Sumaedi, 2013).

Within the SE domain, the concept of loyalty has an ambivalent configuration. Some scholars highlight how consumers turn to SE to maximize the associated social benefits and thus find new individuals to connect with by turning to different platforms (Milanova & Maas, 2017). SE consumers may also prefer not to turn continually to the same community. Maintaining weak ties and relating to networks of strangers is seen as a way of decreasing vulnerability and stricter control over the group (Benkler, 2004). On the other hand, several studies reveal how consumers in SE give great weight to utilitarian reasons, making them more likely to keep using the same services and platforms when satisfied (Jiang & Lau, 2021). It is interesting to note that, rather than manifesting itself as the ultimate effect, loyalty in SE is built up along the whole customer journey, as each stage of consumption is essential in contributing to it. As illustrated by Akhmedova et al. (2020), in the pre-purchase phase the website and app design acts as a strategic antecedent of loyalty by facilitating the process of searching, comparing, negotiating, and bargaining, and by impacting on quality perception. In the purchase phase, a key role is played by platform responsiveness and reliability in dealing with customer issues and mistakes that need attention to increase loyalty. Finally, in the post-purchase phase, loyalty largely depends on interaction with the peer service provider, whose professionalism and skills need to be overseen by the platform. The authors point out that SE platforms need to control at least two of these elements to enhance loyalty; it requires multiple achievements to implement a successful strategy. Furthermore, not all these elements appeal equally to every consumer: male individuals aged below 35 are more sensitive to the efficiency of websites that facilitate knowledge-based trust through interaction (Akhmedova et al., 2020).

Consistent with the nature of SE, relational aspects are important in supporting loyalty. Consumers engaging with peer providers and platforms may derive positive outcomes associated with maintaining a relationship in the long term. Yang et al. (2017) show how consumers perceive confidence, social benefits, and safety benefits as having a very strong effect on loyalty. This means that platforms capable of reducing anxiety, which facilitate contact between parties and which are perceived as secure and reliable, are more likely to develop loyal relations with consumers. Interestingly, their study shows no link between special treatment and loyalty, as is the case with traditional services. While non-SE services might increase loyalty through faster services, monetary savings, or the customized additional services that are often provided in relationship marketing programmes (Morgan et al., 2000), these elements do not appeal to SE consumers. As it is, the very nature of SE provides highly personalized services to consumers based on their specific needs. Thus, as these special treatments are structurally part of SE services, consumers do not value them as additional benefits, and they do not impact on their loyalty intention (Yang et al., 2017). The importance of relational aspects for loyalty has been widely highlighted in both general and sectoral investigations. Marimon et al. (2021) separately investigate the contribution of consumers–providers and consumers–platform relations to loyalty, demonstrating how both agents are equally responsible for the final outcome. On the other side of the coin, negative experiences with one of the two imply an overall decrease of loyalty intentions. Particularly when the negative experience involves the relationship with the provider, consumers perceive a decrease in trust that is transferred in a non-hierarchical way to the platform (Grüner et al., 2024). A similar aspect is evidenced in the second-hand market, where loyalty to the platform extends to the specific brands provided (Abbes et al., 2020).

Relational aspects become very important in the context of service failure. As in SE, consumers see the provider as a peer. They are more likely to appreciate the personal touch in the recovery interaction, rather than monetary compensation or procedural efficacy in solving the problem, as is the case with traditional business services (Lu et al., 2020; Shuqair et al., 2021).

As to cases of service quality and satisfaction, research contributions on loyalty in SE are very often focused on a specific industry, with a view to identifying those elements impacting on loyalty that are due to specific products or services. For example, in the accommodation industry, the likelihood travellers will repeat purchase is increased by the hosts' acceptance rate and listing capacity (Xie et al., 2019), hedonic value (Lee & Kim, 2018), expectations of a unique experience (Mao & Lyu, 2017), frequency of past stays (Xie et al., 2019), and utility and familiarity (Möhlmann, 2015). Focusing on Airbnb, Tajeddini et al. (2022) distinguish revisit intention from customer loyalty, studying how not all the elements that determine the former necessarily impact on the latter. In their research, they conceptualize social value as consumers' desire for recognition, approval, and acceptance among their peers, the hosts, and the neighbouring community. Surprisingly, what they identify is that social value has a significant role in fostering revisit intention but has no effect on creating loyalty in the long run. Conversely, the mood-enhancement effect coming from finding a cost-effective solution is positively and significantly associated with loyalty.

In the mobility industry, Hou et al. (2021) explore the impact of switching costs and trust on loyalty in the ride-sharing services, pointing to the positive effect of implicit costs (e.g. waiting costs, the extra cost of a detour, and the cost of abandoning other alternatives). Lee and Wong (2021) base their research on attitudinal loyalty and confirm the positive effect of WOM on loyalty. They also indicate the role of environmental awareness, price consciousness, perceived usefulness, ease of use, safety, and value as key predictors of WOM in ride-hailing. Möhlmann (2015) focuses on B2C car-sharing services (e.g. Car2go), showing the positive impact of community belonging and utility on loyalty. Surprisingly, she provides evidence that, for this specific service, satisfaction has no role in the likelihood of choosing that sharing option again.

Case Study: Managing Consumers Through Revenue Model Innovation: Examples from Food Sharing Platforms[1]

Among the mechanisms by which consumer retention may be fostered, reconfigurations of value capture modes are recurrent. Value capture is one of the three dimensions of the business model (along with value creation and delivery) and refers to the earning logic of the business. This concept thus covers the revenue model (e.g. subscription fees or pay-per-use), the pricing strategy (e.g. fixed or differentiated prices), and the paying side (e.g. consumers, providers, or third parties). Grieco and Iasevoli (2022) have investigated the ways in which SE platforms manage their value capture mechanisms through a longitudinal analysis, establishing that, over time, platforms moved from one option to another with a view to achieving wider impact and increased consumer lock-in.

Food sharing platforms represent a specific configuration by dint of being hybrid, merging economic and social/environmental missions. The revenue model is evidently a critical element in the platform business model, and its innovation often reflects the reconfiguration of other elements, mostly the offer and the target segments. In the cases analysed, innovation of this element concerned different aspects, such as pricing strategies (e.g. differentiation in delivery costs), the paying side (e.g. moving from charging providers to charging end users), and the revenue stream itself (e.g. from pay-per-use to subscription). Interestingly, the ways these platforms innovate value capture mechanisms over time reveal the different aims its management identifies with.

Attracting consumers: This need is particularly strong with food sharing platforms and SE platforms in general, primarily in the first phase after platform launch. Winning customers is essential when entering the market and fostering awareness raising is strategic. An example comes from Last-minute-sotto-casa, an Italian platform founded in 2017 in Turin to allow supermarkets to digitally manage the expiry dates of their products and send real-time offers to consumers living close to the shops. Two years after its launch, the platform

[1] Data have been collected through semi-structured interviews to platforms' owners and managers of six food sharing platforms (Olio, Regusto, Babaco Market, Best Before, Last-minute-sotto-casa, Bring the food). Collected data have been integrated with a desk research on platforms' websites, press releases, and reports.

changed the category of providers, switching from bulk traders to supermarkets, which were thought to be facing a higher risk of food waste. In this new business model configuration, the choice was made to allow consumers to use the app free of charge while charging providers a subscription fee to be present on the platform. In this way, the likelihood of trying the platform could be increased for consumers who had not used it before or were not particularly waste-conscious.

Reaching different consumer segments: The possibility of reaching different segments is pivotal for business growth. Offering innovation allows this goal to be met and adapting the revenue model is a potential strategy for achieving success. The Olio platform provides an interesting example. It was launched in 2015 in the UK to connect individuals with food for donation with those in the same neighbourhood in need of it. Although starting in a very informal way, the platform quickly achieved great success and today it is available in 49 different countries all over the world with more than seven million users. One goal of the platform was to expand its user base and to this end Olio moved in two directions: allowing individuals to offer both food and non-food items and giving the option of selling items along with making donations. In this way, the platform was opened up to more individuals and to traders. From the revenue model point of view, a winning solution was to adopt a freemium model, which allows users who are unwilling to pay to continue using the app free of charge. At the same time, those who for several reasons are more interested can pay a subscription and benefit from additional services on offer.

Customizing the offer: The experiential side is often lacking in this kind of platform where the key motivational drivers tend to be the economic benefits consumers derive from buying discounted food that would otherwise be wasted, or the environmental awareness of doing something good while combatting food waste. Yet food sharing platforms have started to work on the consumer platform experience to increase consumer engagement. In this context, value capture innovation has followed the adoption of new and more personalized offerings. An example comes from Babaco Market, launched in 2020 to offer fruit and vegetable boxes through subscription. The platform concentrates on products that are difficult to sell through traditional retail outlets; aesthetic defects or surplus production mean they would otherwise remain unsold. A key change to the platform's business model was to introduce one-off orders consumers can make apart from their regular boxes, with the opportunity for consumers to create their own box by choosing from a wide range of products. The goal of this change was to increase the personalization of the offer, and value capture was consistently innovated by introducing differentiated delivery costs made on the basis of city location.

Fostering consumer loyalty: As the ultimate goal of companies is to have loyal consumers, fostering user retention is a key aspect in managing consumers. As mentioned above, this could be harder for SE platforms due to their transaction-centred nature which hampers the creation of switching costs and lock-in strategies. In this respect, value capture mechanisms may be relevant for platforms in a position to both adopt recurring revenue streams which have a long-term perspective but are also utility unbounded (e.g. subscription) and integrate more streams, thereby widening earning potential. An example of this comes from Best Before, founded in 2020 to recover food production surpluses by selling imperfect products at discounted prices. Changes made to the platform over time also satisfy the need to increase consumer engagement and loyalty. Specifically, the platform added food-saving boxes whose contents are unknown to consumers except for a few hints. These boxes were sold regularly via subscription fees, thereby enabling a long-term relationship with consumers and promoting diversification away from the one-shot transactional model.

Attracting providers: While winning and retaining consumers is essential for platform survival, food sharing platforms, as SE platforms, owe their success to the direct and indirect network effects they generate by looking after both sides equally. Thus, the benefit to consumers is enhanced by attracting a suitable number of providers, given value creation depends on them. Value capture innovation could also be managed with this goal of furthering provider participation. An example comes from Regusto, launched in 2016 as both a B2C and B2B platform connecting different entities (e.g. government, NGOs, businesses, and individuals) as providers and consumers. When the COVID-19 pandemic broke out, the platform changed its business model and started focusing solely on the B2B side (business providers donating food to non-profit organizations), turning to blockchain technology to accurately track and communicate companies' commitment to sustainability. In this new version, the choice was to leave the platform free of charge for non-profit organizations and to charge business providers a transaction fee for each kg of products donated through the platform. At first it was a challenge to get them on board, given their reluctance to adapt to new systems and tools. This obstacle has been overcome by offering providers the opportunity of buying more services, for example, a certified impact report, strategically important for those companies interested in CSR activities and disclosure.

4.4 Does Sharing Economy Imply a Regime Shift in Consumption Patterns?

SE implies a mindset shift in consumption patterns. The underlying principle is expressed well by Belk's (2007) "why not share rather than own?" question, which synthetizes the main effect SE platforms are having on consumer behaviour. The possibility of accessing assets instead of owning them allows individuals to shed the burden of possessing something and leads to questioning the actual need to purchase goods. This is particularly clear when focusing on the narrow interpretation of SE, but it could also be extended to the wide interpretation. Even when ownership transfer is envisaged rather than access (e.g. redistribution, reselling, gift-giving), there is a different consumption pattern that takes consumers away from traditional B2C businesses. This has led scholars to wonder whether SE has the potential to cause a shift in the consumption regime. They are therefore conducting empirical research into understanding the impact SE platforms have on traditional businesses, not only in terms of business model adaptation but also with reference to consumer behaviour (Fahmy, 2020; Fang & Li, 2022). Their goal is to assess whether the new consumption mode promoted by SE could be the condition for a definitive regime switch from ownership to access (Grieco & Palagonia, 2024). These studies go a step further, compared to those mentioned above, by investigating similar sets of variables, yet they go beyond the boundaries of the purchase experience (i.e. evaluation of service quality, assessment of satisfaction, continuance intention, and loyalty behaviour). Most studies into consumer behaviour in SE are focused on access acceptance, understanding the motives that drive individuals towards SE (Baumeister & Wangenheim, 2014). Yet poor attention has been given to the effect this has in the long run, and on the extent to which SE becomes a permanent choice in consuming specific products or services (Barbu et al., 2018). This regime shift can be attributed to different factors, such as growing concerns about climate change (Bostman & Rogers, 2010; Albinsson & Yasanthi Perera, 2012) and economical and institutional reasons (Eckhardt et al., 2010).

At a general level, the development of digital technologies is partly driving this regime shift, with the growth of e-commerce already causing a shift from traditional purchasing habits. Fahmy (2020) has investigated this aspect by studying the proportion of e-commerce retail sales to personal consumption in the USA over a 20-year period (1999–2019). The author concludes that the regime shift caused by SE is clearly evidenced and runs in parallel with the diffusion of digital technologies. Another important aspect is that the regime shift is usually linked to satisfaction with SE services and the likelihood of choosing those SE services again in the future. These elements are often investigated as key determinants of shifting behaviour across different industries (Barbu et al., 2018). However, there may be exceptions linked to consumer characteristics or contextual factors. Santoso et al. (2018) focus on car-sharing services to understand the impact of economic benefits, trust, and satisfaction with SE on willingness to buy a new car. Their findings highlight different considerations between those who already own a private car and those who do not. The first group does not exhibit any shift from ownership intention due to their satisfaction with the car-sharing service. What inhibits their buying a second car is an evaluation of the economic benefits derived from access rather than private possession. Conversely, satisfaction determines a regime shift for non-owners, who are less interested in buying a car if they are happy with the shared solution (Santoso et al., 2018).

Fang and Li (2022) claim that regime shift could follow from two sets of factors: on the one hand, from the features of the traditional business (origin) pushing consumers to look for better alternatives; on the other hand, from SE platforms (destination) and their characteristics attracting consumers away from their established consumption patterns. Following this distinction, in their research the authors apply the push-pull-mooring theory to the accommodation industry in order to understand the variables determining consumers' shift from a third-party online travel agent platform to Airbnb. They establish that the regime shift in this domain is caused by both origin and destination: in the former, low structural assistance and low standardization push consumers to Airbnb; in the latter, consumers are attracted by the monetary savings, sustainability, interactivity, and authentic experience available through the SE platform. A key role is played by the degree of substitutability: the higher it is perceived to be, the greater will be the tendency to shift from traditional services to Airbnb. However, network externalities do not play a role in influencing shifting behaviour (Fang & Li, 2022).

Two important considerations should be taken into account with regime shift. The first is that, at a wider level, shifting behaviour mostly refers to partial replacement, with consumers adopting new patterns of consumption that run alongside preexisting ones (Ye & Potter, 2011). Despite the notable growth in SE platforms, it is more likely that they will continue to offer an additional option to traditional purchases. This is mainly because, according to the endowment effect, possession still plays a role in determining the value of an object (Fota et al., 2019). A complete substitution of traditional businesses with SE services may occur within a defined time frame; for example, car-sharing might be a good solution for a young adult unable to afford buying a new one (Magnani & Re, 2020). It may also apply to a specific product or service category (e.g. not buying a new car because of the availability of car-sharing services) (Santoso et al., 2018; Fang & Li, 2022). The second consideration is that regime shift cannot be studied in SE as a whole, rather through sector-specific investigations that account for important differences in terms of price, durability, visibility of consumption, and main purchase goal, associated with different categories of products and services (Baumeister & Wangenheim, 2014).

Insights: Satisfaction, Repurchase, and Shift Intention in the Mobility Industry

The following figures show the results of an empirical investigation carried out on mobility sharing consumers. A questionnaire was sent out through the Google Form platform and self-administered by a sample of 399 users.

Satisfaction and continuance intention

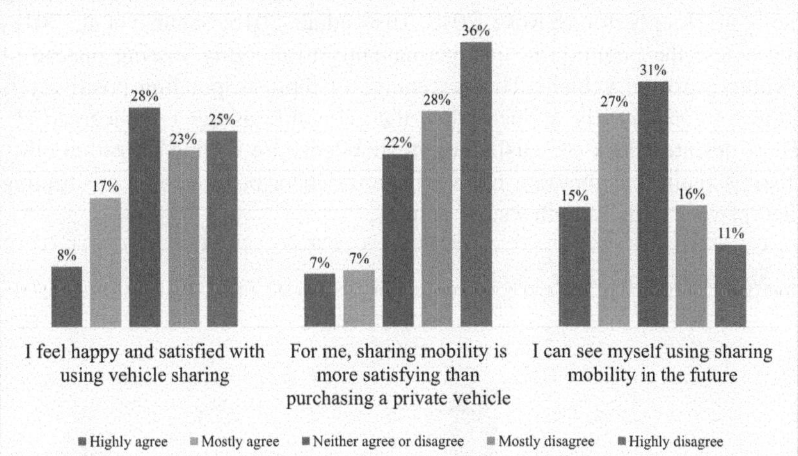

It is worth noting how nearly half of the sample said they were dissatisfied by the mobility sharing service they used (23% mostly disagree and 25% completely disagree), and only 14% of respondents consider mobility sharing to give more satisfaction than owning a private vehicle. Yet, the configuration changes when asked about continuance intention, as only 11% state their intention not to use mobility sharing in the future.

Shift intention (1/2)

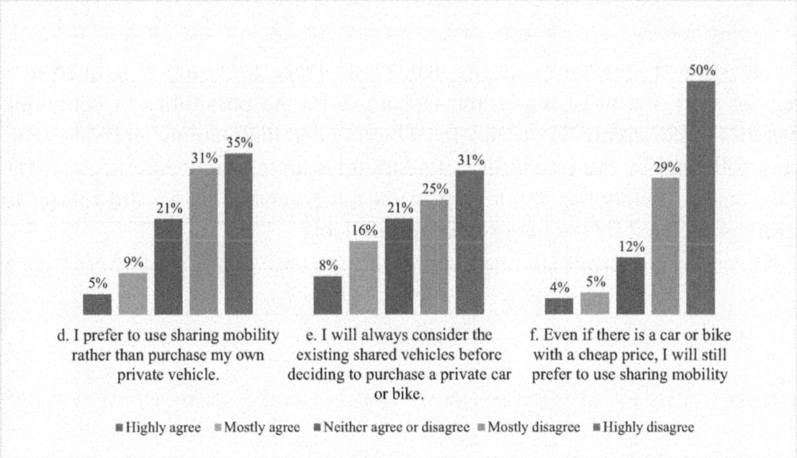

The questions about shift intention have been developed based on existing literature (Lamberton & Rose, 2012; Tussyadiah, 2016; Santoso et al., 2018) and assess the intention to turn permanently to mobility sharing instead of owning a private vehicle. The percentage of those responding positively is quite low and mostly applies when the economic aspect is concerned (f). Thus, despite a lack of satisfaction, respondents are willing to use mobility sharing again, but they are not very interested in permanently substituting their private vehicles with shared options.

Shift intention (2/2)

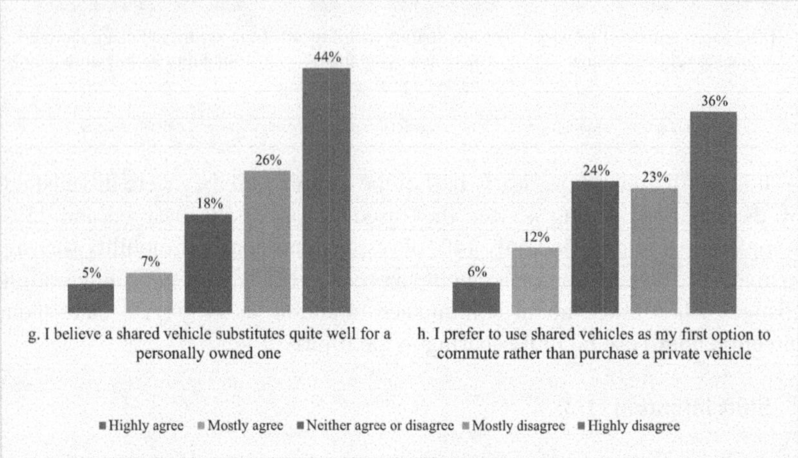

Questioned about the reasons that would make them more inclined to a regime shift, the most important appears to be the possibility of economic benefit (i) (i.e. saving the money from buying and maintaining a private vehicle), followed by the possibility of adopting a more sustainable lifestyle (l). Conversely, community belonging (g) is not perceived as a valid reason to shift, as declared by the majority of respondents.

Reasons to shift

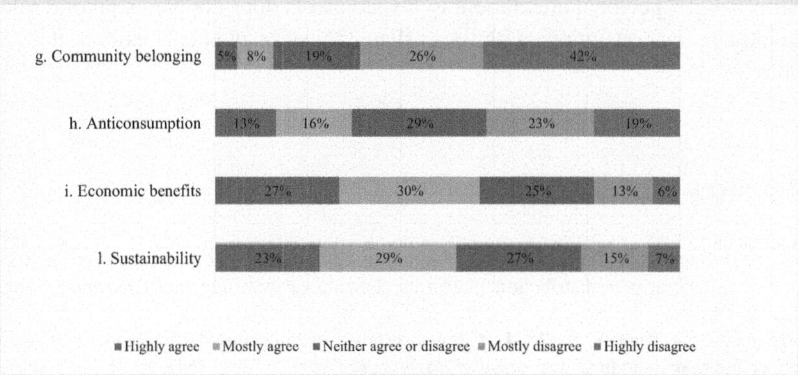

Barriers to shift

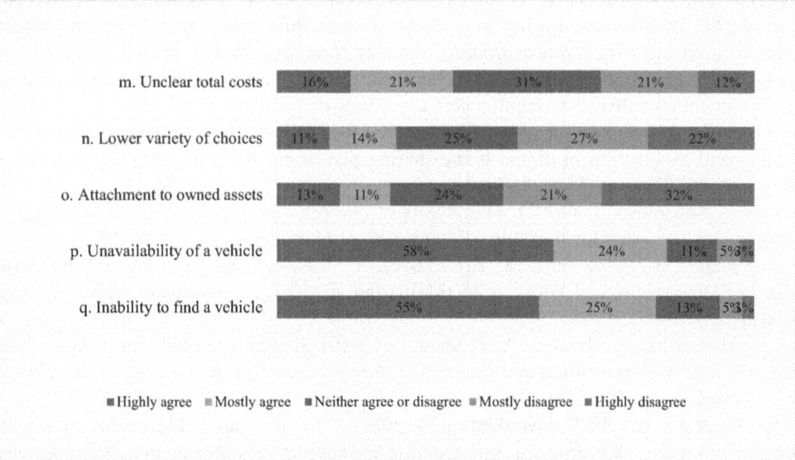

A final consideration concerns the reasons why respondents would not permanently shift to a shared vehicle. The barriers are mostly linked to functional aspects such as the risk of experiencing difficulties in finding a suitable vehicle when needed (p) and the fear of lacking the competence to find one at all (q). More than half of the sample strongly agree with these as the key elements preventing them from considering a shift towards shared mobility. Surprisingly, personal attachment to owned assets does not play a pivotal role in influencing consumers, with fewer than 25% of respondents naming this as an issue.

References

Abbes, I., Hallem, Y., & Taga, N. (2020). Second-hand shopping and brand loyalty: The role of online collaborative redistribution platforms. *Journal of Retailing and Consumer Services, 52*, 101885.

Ackermann, L., & Tunn, V. S. (2024). Careless product use in access-based services: A rebound effect and how to address it. *Journal of Business Research, 177*, 114643.

Ahrholdt, D. C., Gudergan, S. P., & Ringle, C. M. (2017). Enhancing service loyalty: The roles of delight, satisfaction, and service quality. *Journal of Travel Research, 56*(4), 436–450.

Akhmedova, A., Marimon, F., & Mas-Machuca, M. (2020). Winning strategies for customer loyalty in the sharing economy: A mixed-methods study. *Journal of Business Research, 112*, 33–44.

Albinsson, P. A., & Yasanthi Perera, B. (2012). Alternative marketplaces in the 21st century: Building community through sharing events. *Journal of Consumer Behaviour, 11*(4), 303–315.

Al-dweeri, R. M., Obeidat, Z. M., Al-dwiry, M. A., Alshurideh, M. T., & Alhorani, A. M. (2017). The impact of e-service quality and e-loyalty on online shopping: Moderating effect of e-satisfaction and e-trust. *International Journal of Marketing Studies, 9*(2), 92–103.

Amat-Lefort, N., Marimon, F., & Mas-Machuca, M. (2020). Towards a new model to understand quality in collaborative consumption services. *Journal of Cleaner Production, 266*, 121855.

Amat-Lefort, N., Marimon, F., & Mas-Machuca, M. (2023). Guest and host perspectives of service quality and satisfaction in digital home-sharing platforms. *Journal of Quality Assurance in Hospitality & Tourism, 24*(6), 859–884.

Anatolevena Anisimova, T. (2007). The effects of corporate brand attributes on attitudinal and behavioural consumer loyalty. *Journal of Consumer Marketing, 24*(7), 395–405.

Arteaga-Sánchez, R., Belda-Ruiz, M., Ros-Galvez, A., & Rosa-Garcia, A. (2020). Why continue sharing: Determinants of behavior in ridesharing services. *International Journal of Market Research, 62*(6), 725–742.

Azmi, A., Ahmad, M. A., Abdullah, A., & Shaw, G. (2019). Shopping in Malaysia: A study of international tourists' expectation and experience. *International Journal of Innovation, Creativity and Change, 8*(4), 199–212.

Bahadur, W., & Ali, A. (2023). Investigating the effect of service quality dimensions on travellers' satisfaction with couchsurfing accommodation and subjective wellbeing in a sharing economy. *Economic research-Ekonomska istraživanja, 36*(3).

Bakti, I. G. M. Y., & Sumaedi, S. (2013). An analysis of library customer loyalty: The role of service quality and customer satisfaction, a case study in Indonesia. *Library Management, 34*(6/7), 397–414.

Barbu, C. M., Florea, D. L., Ogarcă, R. F., & Barbu, M. C. (2018). From ownership to access: How the sharing economy is changing the consumer behavior. *Amfiteatru Economic, 20*(48), 373–387.

Baumeister, C., & Wangenheim, F. V. (2014). Access vs. ownership: Understanding consumers' consumption mode preference. *Ownership: Understanding consumers' consumption mode preference* (July 7, 2014).

Belk, R. (2007). Why not share rather than own? *The Annals of the American Academy of Political and Social Science, 611*(1), 126–140.

Benkler, Y. (2004). Sharing nicely: On shareable goods and the emergence of sharing as a modality of economic production. *Yale Lj, 114*, 273.

Berg, L., Slettemeås, D., Kjørstad, I., & Rosenberg, T. G. (2020). Trust and the don't-want-to-complain bias in peer-to-peer platform markets. *International Journal of Consumer Studies, 44*(3), 220–231.

Bhattacherjee, A. (2001a). An empirical analysis of the antecedents of electronic commerce service continuance. *Decision Support Systems, 32*(2), 201–214.

Bhattacherjee, A. (2001b). Understanding information systems continuance: An expectation-confirmation model. *MIS Quarterly*, 351–370.

Bolton, R. N., & Drew, J. H. (1991). A multistage model of customers' assessments of service quality and value. *Journal of Consumer Research, 17*(4), 375–384.

Bostman, R., & Rogers, R. (2010). *What is mine is yours: The rise of collaborative consumption*. HarperCollins.

Brady, M. K., & Cronin, J. J., Jr. (2001). Some new thoughts on conceptualizing perceived service quality: A hierarchical approach. *Journal of Marketing, 65*(3), 34–49.

Chen, M. K., & Sheldon, M. (2015). *Dynamic pricing in a labor market: Surge pricing and the supply of Uber driver-partners*. University of California (Los Angeles) Working Paper. URL http://citeseerx.ist.psu.edu/viewdoc/download

Cheng, X., Fu, S., & De Vreede, G. J. (2018). A mixed method investigation of sharing economy driven car-hailing services: Online and offline perspectives. *International Journal of Information Management, 41*, 57–64.

Cho, Y. (2020). Exploring determinants of performance indicator and customer satisfaction of accommodation sharing. *The Journal of Asian Finance, Economics and Business, 7*(3), 201–210.

Chu, J., & Manchanda, P. (2016). Quantifying cross and direct network effects in online consumer-to-consumer platforms. *Marketing Science, 35*(6), 870–893.

Cronin, J., & Taylor, S. A. (1992). Measuring service quality: Are-examination and extension. *Journal of Marketing, 56*(3), 55–68.

Cronin, J., & Taylor, S. A. (1994). SERVPERF versus SERVQUAL: Reconciling performance-based and perceptions-minus-expectations measurements of service quality. *Journal of Marketing, 58*(1), 125–131.

Dabholkar, P. A., Thorpe, D. I., & Rentz, J. O. (1996). A measure of service quality for retail stores: Scale development and validation. *Journal of the Academy of Marketing Science, 24*, 3–16.

Dütschke, E., Frondel, M., Schleich, J., & Vance, C. (2018). Moral licensing—another source of rebound? *Frontiers in Energy Research, 6*, 38.

Eckhardt, G. M., Belk, R., & Devinney, T. M. (2010). Why don't consumers consume ethically? *Journal of Consumer Behaviour, 9*(6), 426–436.

Ertz, M., Boily, É., Sun, S., & Sarigöllü, E. (2022). Role transitions at the prosumer level: Spillover effects in the collaborative economy from an interactive marketing perspective. *European Journal of Marketing, 56*(10), 2721–2748.

European Commission. (2017). *Exploratory study of consumer issues in online peer-to-peer platform markets (Final Report)*. DG Justice and Consumers, European Union.

Fahmy, H. (2020). Is the sharing economy causing a regime switch in consumption? *Journal of Applied Economics, 23*(1), 281–298.

Fang, Y. H., & Li, C. Y. (2022). Does the sharing economy change conventional consumption modes? *International Journal of Information Management, 67*, 102552.

Fota, A., Wagner, K., & Schramm-Klein, H. (2019). Is renting the new buying? A quantitative investigation of the determinants of the rental-commerce intention. *The International Review of Retail, Distribution and Consumer Research, 29*(5), 582–599.

Ghazali, E., Nguyen, B., Mutum, D. S., & Mohd-Any, A. A. (2016). Constructing online switching barriers: examining the effects of switching costs and alternative attractiveness on e-store loyalty in online pure-play retailers. *Electronic Markets, 26*, 157–171.

Grieco, C., & Iasevoli, G. (2022). Paths and patterns of value capture innovation in sharing economy. *Journal of Revenue and Pricing Management, 21*(3), 255–261.

Grieco, C., & Palagonia, C. (2024). Delving into the behaviour of sharing economy consumers: A literature review. *Journal of Consumer Marketing, 41*(2), 162–179.

Grönroos, C. (1982). An applied service marketing theory. *European Journal of Marketing, 16*(7), 30–41.

Grüner, A., Postel, L., & Schumann, J. H. (2024). Sharing is caring? The effect of negative peer-to-peer experiences on loyalty intentions in the sharing economy. *Journal of Business Research, 181*, 114757.

Hawkins, K., & Vel, P. (2013). Attitudinal loyalty, behavioural loyalty and social media: An introspection. *The Marketing Review, 13*(2), 125–141.

Haywood-Farmer, J. (1988). A conceptual model of service quality. *International journal of operations & production management,* 8(6), 19–29.

Hess, R. L., Jr., Ganesan, S., & Klein, N. M. (2003). Service failure and recovery: The impact of relationship factors on customer satisfaction. *Journal of the Academy of Marketing Science, 31*(2), 127–145.

Hou, T., Cheng, X., & Cheng, X. (2021). The role of transaction cost and trust in e-loyalty: A mixed-methods study of ride-sharing. *Information Technology & People, 34*(3), 1018–1038.

Hughes, R., & MacKenzie, D. (2016). Transportation network company wait times in Greater Seattle, and relationship to socioeconomic indicators. *Journal of Transport Geography, 56*, 36–44.

Jiang, Y., & Lau, A. K. (2021). Roles of consumer trust and risks on continuance intention in the sharing economy: An empirical investigation. *Electronic Commerce Research and Applications, 47*, 101050.

Jin, R., & Chen, K. (2020). Impact of value cocreation on customer satisfaction and loyalty of online car-hailing services. *Journal of Theoretical and Applied Electronic Commerce Research, 16*(3), 432–444.

Kaur, S., Tan, N., & Dutta, M. J. (2016). Media, migration and politics: The coverage of the little India riot in The Straits Times in Singapore. *Journal of Creative Communications, 11*(1), 27–43.

Kim, D. J., Ferrin, D. L., & Rao, H. R. (2009). Trust and satisfaction, two stepping stones for successful e-commerce relationships: A longitudinal exploration. *Information Systems Research, 20*(2), 237–257.

Knutson, B., Stevens, P., Wullaert, C., Patton, M., & Yokoyama, F. (1990). LODGSERV: A service quality index for the lodging industry. *Hospitality Research Journal, 14*(2), 277–284.

Kristensen, K., Martensen, A., & Gronholdt, L. (1999). Measuring the impact of buying behaviour on customer satisfaction. *Total Quality Management, 10*(4-5), 602–614.

Kumar, V. I. S. W. A. N. A. T. H. A. N., & Shah, D. (2004). Building and sustaining profitable customer loyalty for the 21st century. *Journal of Retailing, 80*(4), 317–329.

Kumar, A., Gupta, A., Parida, M., & Chauhan, V. (2022). Service quality assessment of ride-sourcing services: A distinction between ride-hailing and ride-sharing services. *Transport Policy, 127*, 61–79.

Lamberton, C. P., & Rose, R. L. (2012). When is ours better than mine? A framework for understanding and altering participation in commercial sharing systems. *Journal of marketing*, 76(4), 109–125.

Lee, S., & Kim, D. Y. (2018). The effect of hedonic and utilitarian values on satisfaction and loyalty of Airbnb users. *International Journal of Contemporary Hospitality Management, 30*(3), 1332–1351.

Lee, C. K. H., & Wong, A. O. M. (2021). Antecedents of consumer loyalty in ride-hailing. *Transportation Research Part F: Traffic Psychology and Behaviour, 80*, 14–33.

Liang, L. J., Choi, H. C., & Joppe, M. (2018). Exploring the relationship between satisfaction, trust and switching intention, repurchase intention in the context of Airbnb. *International Journal of Hospitality Management, 69*, 41–48.

Lin, C. S., Wu, S., & Tsai, R. J. (2005). Integrating perceived playfulness into expectation-confirmation model for web portal context. *Information and Management, 42*(5), 683–693.

Loiacono, E. T., Watson, R. T., & Goodhue, D. L. (2002). WebQual: A measure of website quality. *Marketing Theory and Applications, 13*(3), 432–438.

Lu, L., Cai, R., & King, C. (2020). Building trust through a personal touch: Consumer response to service failure and recovery of home-sharing. *Journal of Business Research, 117*, 99–111.

Magnani, G., & Re, B. (2020). Lived experiences about car sharing in young adults: Emerging paradoxes. *Italian Journal of Marketing, 2020*(2), 207–229.

Mao, Z., & Lyu, J. (2017). Why travelers use Airbnb again? An integrative approach to understanding travelers' repurchase intention. *International Journal of Contemporary Hospitality Management, 29*(9), 2464–2482.

Marimon, F., Llach, J., Alonso-Almeida, M., & Mas-Machuca, M. (2019). CC-Qual: A holistic scale to assess customer perceptions of service quality of collaborative consumption services. *International Journal of Information Management, 49*, 130–141.

Marimon, F., Mas-Machuca, M., & Llach, J. (2021). Making the long and adventurous journey from quality to loyalty. *Total Quality Management and Business Excellence, 32*(15-16), 1713–1731.

Merino-Saum, A., Jemio, P. R., Hansmann, R., & Binder, C. R. (2023). Drivers and barriers to participation in the sharing economy: Does the environment really matter? A systematic review of 175 scientific articles. *Resources, Conservation and Recycling, 198*, 107121.

Meshulam, T., Font-Vivanco, D., Blass, V., & Makov, T. (2023). Sharing economy rebound: The case of peer-to-peer sharing of food waste. *Journal of Industrial Ecology, 27*(3), 882–895.

Milanova, V., & Maas, P. (2017). Sharing intangibles: Uncovering individual motives for engagement in a sharing service setting. *Journal of Business Research, 75*, 159–171.

Möhlmann, M. (2015). Collaborative consumption: determinants of satisfaction and the likelihood of using a sharing economy option again. *Journal of Consumer Behaviour, 14*(3), 193–207.

Morgan, R. M., Crutchfield, T. N., & Lacey, R. (2000). Patronage and loyalty strategies: Understanding the behavioral and attitudinal outcomes of customer retention programs. *Relationship marketing: Gaining competitive advantage through customer satisfaction and customer retention* (pp. 71–87).

Nozari, M. A., Ghadikolaei, A. S., Govindan, K., & Akbari, V. (2021). Analysis of the sharing economy effect on sustainability in the transportation sector using fuzzy cognitive mapping. *Journal of Cleaner Production, 311*, 127331.

OECD. (2017). *Trust in peer platform markets: Consumer survey findings* (OECD Digital Economy Papers No. 263). OECD.

Oliver, R. L. (1980). A cognitive model of the antecedents and consequences of satisfaction decisions. *Journal of Marketing Research, 17*(4), 460–469.

Oliver, R. L. (2014). *Satisfaction: A behavioral perspective on the consumer: A behavioral perspective on the consumer*. Routledge.

Parasuraman, A., Zeithaml, V. A., & Berry, L. L. (1985). A conceptual model of service quality and its implications for future research. *Journal of Marketing, 49*(4), 41–50.

Parasuraman, A., Zeithaml, V. A., & Berry, L. L. (1988). Servqual: A multiple-item scale for measuring consumer perc. *Journal of Retailing, 64*(1), 12.

Parasuraman, A., Zeithaml, V. A., & Malhotra, A. (2005). ES-QUAL: A multiple-item scale for assessing electronic service quality. *Journal of Service Research, 7*(3), 213–233.

Philip, G., & Hazlett, S. A. (1997). The measurement of service quality: A new P-C-P attributes model. *International Journal of Quality and Reliability Management, 14*(3), 260–286.

Phuong, N. N., & Trang, T. T. (2018). Repurchase intention: The effect of service quality, system quality, information quality, and customer satisfaction as mediating role: A PLS approach of m-commerce ride hailing service in Vietnam. *Marketing and Branding Research, 5*, 78–91.

Priporas, C. V., Stylos, N., Rahimi, R., & Vedanthachari, L. N. (2017). Unraveling the diverse nature of service quality in a sharing economy: A social exchange theory perspective of Airbnb accommodation. *International Journal of Contemporary Hospitality Management, 29*(9), 2279–2301.

Rasheed, A. F., & Nandukrishna, A. T. (2023). Caring the sharing apps–exploration of the factors affecting user satisfaction with sharing economy apps. *Journal of Global Marketing, 36*(3), 225–239.

Santoso, A. S., Hadiansah, I., & Christoni, E. (2018). Does satisfaction of P2P online transportation affect consumer willingness to purchase their own vehicles? *International Journal of Business and Globalisation, 21*(3), 408–425.

Shao, Z., Li, X., Guo, Y., & Zhang, L. (2020). Influence of service quality in sharing economy: Understanding customers' continuance intention of bicycle sharing. *Electronic Commerce Research and Applications, 40*, 100944.

Shuqair, S., Pinto, D. C., So, K. K. F., Rita, P., & Mattila, A. S. (2021). A pathway to consumer forgiveness in the sharing economy: The role of relationship norms. *International Journal of Hospitality Management, 98*, 103041.

Slack, N., Singh, G., & Sharma, S. (2020). The effect of supermarket service quality dimensions and customer satisfaction on customer loyalty and disloyalty dimensions. *International Journal of Quality and Service Sciences, 12*(3), 297–318.

Stauss, B., & Weinlich, B. (1997). Process-oriented measurement of service quality: Applying the sequential incident technique. *European Journal of Marketing, 31*(1), 33–55.

Stevens, P., Knutson, B., & Patton, M. (1995). DINESERV: A tool for measuring service quality in restaurants. *Cornell Hotel and Restaurant Administration Quarterly, 36*(2), 56–60.

Stough, R. A., & Carter, E. P. (2023). What was yours is (for now) mine: Prior user knowledge reduces product satisfaction but can improve experiential satisfaction in access-based consumption. *Journal of Consumer Behaviour, 22*(4), 833–847.

Tajeddini, K., Gamage, T. C., Hameed, W. U., Qumsieh-Mussalam, G., Chaijani, M. H., Rasoolimanesh, S. M., & Kallmuenzer, A. (2022). How self-gratification and social values shape revisit intention and customer loyalty of Airbnb customers. *International Journal of Hospitality Management, 100*, 103093.

Tanford, S. (2016). Antecedents and outcomes of hospitality loyalty: A meta-analysis. *Cornell Hospitality Quarterly, 57*(2), 122–137.

Tussyadiah, I. P. (2016). Factors of satisfaction and intention to use peer-to-peer accommodation. *International Journal of Hospitality Management, 55*, 70–80.

Wang, X., Lin, X., & Liu, Z. (2021). Understanding consumers' post-adoption behavior in sharing economy services. *Journal of Computer Information Systems, 61*(3), 275–284.

Wireko-Gyebi, S., Baah, N. G., Nustugbodo, R. Y., Zaazie, P., Senya, E. K., Bosomefi, V., & Amankona, S. (2024). Service quality, satisfaction and loyalty among sharing economy vehicle users. *Cogent Social Sciences, 10*(1), 2327130.

Xie, K. L., Kwok, L., & Wu, J. (2019). Are consumers loyal to home-sharing services? Impacts of host attributes and frequency of past stays. *International Journal of Contemporary Hospitality Management, 31*(3), 1066–1085.

Yang, S., Song, Y., Chen, S., & Xia, X. (2017). Why are customers loyal in sharing-economy services? A relational benefits perspective. *Journal of Services Marketing, 31*(1), 48–62.

Yapp, E. H., & Yeap, J. A. (2023). Assessing the determinants of customer-perceived value and customer satisfaction in e-hailing services: An Importance-Performance Matrix Analysis (IPMA) Approach. *Cogent Business and Management, 10*(1), 2191808.

Ye, C., & Potter, R. (2011). The role of habit in post-adoption switching of personal information technologies: An empirical investigation. *Communications of the Association for Information Systems, 28*(1), 35.

Yoo, B., & Donthu, N. (2001). Developing and validating a multidimensional consumer-based brand equity scale. *Journal of Business Research, 52*(1), 1–14.

Yoo, M., & Yoon, S. (2022). The six lodging attributes that determine travelers' preference on Airbnb or hotel. *Journal of Marketing Analytics, 10*(4), 299–310.

Zervas, G., Proserpio, D., & Byers, J. W. (2021). A first look at online reputation on Airbnb, where every stay is above average. *Marketing Letters, 32*, 1–16.

Zhou, Z., & Zhang, Z. J. (2019). Customer satisfaction of bicycle sharing: Studying perceived service quality with SEM model. *International Journal of Logistics Research and Applications, 22*(5), 437–448.

Ziyad, A., Rehman, Z.-U., Batool, Z., & Khan, A. H. (2020). Influence of service excellence on consumer satisfaction of ridesharing industry. *International Journal for Traffic and Transport Engineering, 10*(4), 468–448.

The manufacturer's authorised representative in the EU is Springer Nature Customer Service Centre GmbH, Europaplatz 3, 69115 Heidelberg, Germany. If you have any concerns regarding our products, please contact ProductSafety@springernature.com

Printed and bound by CPI Group (UK) Ltd, Croydon, CR0 4YY
30/09/2026
02225607-0002